PROGRESS
in
ROBOTICS
and
INTELLIGENT SYSTEMS

VOLUME 2

edited by
George W. Zobrist
and
C.Y. (Pete) Ho

University of Missouri—Rolla

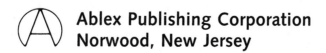

Ablex Publishing Corporation
Norwood, New Jersey

Printed in the United States of America

Library of Congress Cataloging-in-Publication Data
Progress in robotics and intelligent systems.
  Includes bibliographical references.
  1. Robotics.   2. Artificial intelligence.
I. Zobrist, George W. (George Winston), 1934–
II. Ho, C. Y. (Chung You), 1933–
TJ211.P76   1994   670.42′72   89-17746
ISBN 0-89391-593-9 (v. 2)

Ablex Publishing Corporation
355 Chestnut Street
Norwood, New Jersey 07648

Contents

To Kelly—what a pistol!

Series Preface

This series is intended for those individuals involved in robotics and intelligence systems research, design, development, and scholarly activities.

This volume is concerned with state-of-the-art developments in robotics and intelligent systems by providing insight and guidance into specific techniques vital to those concerned with design and implementation of robotics and intelligent system applications.

The material contained in this volume discusses motion learning and vision-based robotics, control algorithms, accuracy issues, networking robots, and robotic programming techniques.

The editors wish to thank the contributing authors for making available the information contained in this book.

George W. Zobrist
C.Y. Ho
University of Missouri-Rolla

May 1994

Experimental Analysis of Convex Volumes Enclosing Parametric Surfaces

Chaman L. Sabharwal
University of Missouri—ROLLA

Thomas G. Melson
Computer Aided Technology, MCAIR

1. INTRODUCTION

Computing intersections between geometric objects in particular surfaces [1] is a key capability of CAD/CAM systems and many other geometric modeling systems. Parametric equations completely separate the roles of independent and dependent variables, both geometrically and algebraically, and allow for any number of variables (i.e., there is a natural extension from two- to three-dimensional space). Parametrically defined objects are inherently bounded because the parameter space is normalized to a unit square. There is no need to carry additional data to define boundaries. The calculation and intersection of bounding volumes for parametric surfaces is used, in a wide range of applications, as a scaffold to mitigate the agony of complex computations. The rectangular parallelepiped bounding volumes are also referred to as bounding boxes, or simply boxes. Bounding boxes are interchangeably used for rectangular parallelepiped bounding volumes. Bounding volumes are useful for parametric surface intersections [2, 3], collision detection [4], and partitioning of polyhedral objects into nonintersecting parts [5].

The intersection between objects and the collision detection problem arises in CAD/CAM applications where one needs to cull disjoint objects before computationally intensive analysis is performed on the objects. Human eyes can detect intersection easily and instantaneously, but it is a difficult problem for the computer to visualize this phenomena. The culling process is done by first enclosing the two objects in bounding volumes and determining whether or not the bounding volumes intersect. The cost of detecting the intersecting bounding volumes is significantly lower than the cost of intersecting the objects. The bounding volumes may intersect, but the objects may not. In that case, a more computation-intensive analysis is performed.

This problem arises also in other areas such as Robotics, where interference detection is to be determined; and Computer Graphics, where hidden surface removal and ray tracing take place. For intersection between surfaces, one subdivides the larger surfaces into smaller and simpler surface pieces. This subdivision process is a selective subdivision in which the surfaces are enclosed in bounding volumes and the bounding volumes are tested for intersection conditions to detect the need for further subdivision of the surfaces. It is easier to detect the intersection condition between the bounding volumes than between the surfaces themselves. Further, in some applications, because the bounding boxes are used for the culling process only, actual intersection between the bounding volumes is not required: Only a flag value is used to indicate the existence or nonexistence of intersection. Note that computation of actual intersection between the bounding volumes leads to a more complex problem of intersection between C^0 surfaces. The designer of an algorithm has two goals at hand: (a) The algorithm should be understandable, easy to code, and easy to debug. (b) It should make optimal use of computer resources with respect to both storage space and execution time. We have these two goals in mind in the analysis of the methods to be considered. Several methods have been used to calculate the bounding volumes for surfaces and to detect the intersection between them. This chapter singles out one method that is mathematically sound, numerically less prone to computational errors, computationally efficient, and easier to understand and implement. The source code and load segments make efficient use of computer resources. An execution time efficiency analysis is performed to determine its suitability for use in production modules.

2.   DISCUSSION OF METHODS

The bounding volumes for the surfaces can be computed in several ways. An important and much-debated issue is how to calculate the enclosing volume. The object of this presentation is to put this issue to rest for a long time. The methods for calculating the bounding volumes for surfaces can be classified as: (a) axis-oriented parallelepipeds [2, 3], (b) surface-oriented parallelepipeds [6], (c) convex hulls [7], (d) ellipsoids [8], and (e) spheroids [9, 10]. There is a need for an ideal method guaranteeing that the surface remains entirely within its bounding volume. However, this task is impossible for an arbitrary parametric surface with no additional structural information. In the absence of additional information, only the sampled points are used to compute the bounding volumes. There is no way to guarantee the behavior of the function values where the surface is not sampled, if samples are the only information available. It is, therefore, desirable to minimize the difference between the computed bounding volume enclosing the surface and the actual geometric volume containing the surface. At the same time, there is a need to strike a balance to achieve the best out of the available methods for bounding volumes.

Analytical comparison shows that methods (b), (c), (d), and (e) are poor choices for one or more of the following reasons: the excessive computation time for the calculation of bounding volumes, the convex hull property of the surfaces, smoothness constraints on surfaces, and the excessive performance time for intersecting the bounding volumes. The torus represents an example of a surface where none of the above-mentioned bounding volume methods works well. The simplicity of a box calculation with axis-oriented parallelepipeds and the associated simplicity in box intersection makes the axis-oriented method more acceptable than other methods.

To keep the comparisons simple, the details of the methods have been simplified. The spirit of the techniques has been retained in order to point out the complexity of the methods. For execution time analysis, run-time tests are performed on the axis-oriented and surface-oriented methods that are currently used at the McDonnell Douglas Corporation. These methods can be characterized as follows: (a) use only the position value, 0-dimensional information, to calculate axis-oriented bounding volumes or (b) use a position value and three direction vectors, 1-dimensional information, to calculate the surface-oriented bounding volumes.

2.1. Details of Axis-Oriented Method

This method is designed to compute approximate bounding boxes, which are oriented along the axes of the coordinate system. Two different methods for computing the axis-oriented bounding volumes, considered here, depend on the nature of the surfaces (e.g., C^0 and C^2 surfaces). However, the method for C^0 surfaces will still be applicable to C^2 surfaces.

2.1.1. Axis-Oriented Method for C^0 Surfaces.

This method is the same as implemented in the original surface/surface intersection algorithm [2, 3]. Since it was used in the surface/surface intersection algorithm as implemented in 1981, it is referred to as SURF81. Since the implementation of this algorithm was revised in 1987 to eliminate the unnecessary calculations and reduce the repeated evaluator calls, the same method in the revised version is referred to as SURF87. In this method, it is easier to calculate the bounding volume and simpler to detect the intersection between the bounding volumes. For computation-intensive practical applications, it is not necessary to have exact bounding volumes because the approximate bounding volumes can be used without loss of information. Sample N^2 points on the surface $R(u, v)$, equally spaced parametrically, calculate the Maximum and Minimum points over the sampled data. It gives an approximate bounding volume for the surface $R(u, v)$. Since the computations are over the sampled points on the surface, there is no guarantee that the surface lies entirely inside the bounding volume. Extensive empirical evidence indicates that, on the average, if the boxes over a surface with D-degree curvature is expanded by 0.D percent, the resulting bounding volumes would normally contain the surface. This is sufficient for practical applications where the numerical accuracy of the bounding volumes is not necessary or required. However, this or any other numerical method can be frustrated by creating sufficiently unrealistic examples. Thus, the bounding volume for a surface $R(u, v)$ is given by two points (Min and Max), such that

$$\text{Min}(i) \leq R(u, v, i) \leq \text{Max}(i)$$

for $i = 1, 2, 3$; $LWU \leq u \leq UPU$ and $LWV \leq v \leq UPV$.

Here LWU and LWV are the lower bounds on the parameters u and v, UPU and UPV are the upper bounds on the parameters u and v, and $i = 1, 2$, and 3 refer to the x-, y-, and z-coordinates of the points.

Determining whether the axis-oriented bounding volumes intersect (or not) is as easy as calculating these volumes. Two bounding volumes (Min_1, Max_1) and (Min_2, Max_2) are disjoint provided there exists an i, $1 \leq i \leq 3$, such that

$$\text{Max}_1(i) < \text{Min}_2(i), \text{ or}$$
$$\text{Max}_2(i) < \text{Min}_1(i).$$

It shows that it is easy to calculate the bounding volumes and it is even easier to detect the intersection condition between the axis-oriented bounding volumes.

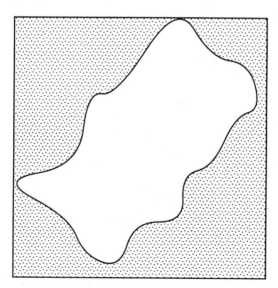

FIGURE 1.1. Axis-Oriented Bounding Box

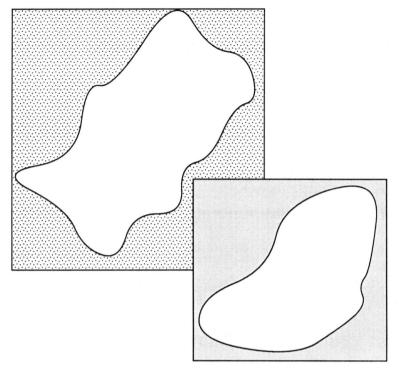

FIGURE 1.2.   Axis-Oriented Boxes Intersect, Surfaces Do Not Intersect

2.1.2.   *Axis-Oriented Method for C² Surfaces.*

This method applies to C^k, $k \geq 2$, surfaces only and was developed at Automation Technology Products [11]. This technique depends on the ability to calculate the second-order partial derivatives of the surface with respect to the surface parameters and the bounds on these derivatives. The discussion of this method warrants the discussion of some terminology. Let (p, q), $(p + dp, q)$, and $(p, q + dq)$ be three points in the parameter space of the surface $R(u, v)$. The surface $R(u, v)$ can be linearly approximated by $L(u, v)$ over the nondegenerate triangle (p, q), $(p + dp, q)$, and $(p, q + dq)$ within a specified tolerance epsilon. The approximation function is given by:

$$L(u, v) = R(p, q) + (u - p)(R(p + dp, q) - R(p, q))/dp$$
$$+ (v - q)(R(p, q + dq) - R(p, q))/dq$$

provided

$$(dp^2 M_1 + 2\ dp\ dq\ M_2 + dq^2 M_3) < 8 * \text{epsilon}$$

where

$$M_1 = \sup\|R_{11}(u, v)\|,$$
$$M_2 = \sup\|R_{12}(u, v)\|,$$
$$M_3 = \sup\|R_{22}(u, v)\|.$$

Here $R_{11}(u, v)$ is the second-order partial derivative with respect to u, $R_{22}(u, v)$ is the second-order partial derivative with respect to v, $R_{12}(u, v)$ is the second-order mixed partial derivative with respect to u and v of $R(u, v)$; *sup* is the maximum value of the two-norms of the second-order derivative function over the specified triangular domain: (p, q), $(p + dp, q)$, and $(p, q + dq)$.

Calculate integers n and m to create a uniform parametric grid, such that

$$(M_1/n^2 + 2M_2/mn + M_3/m^2) < 8 * \text{epsilon}.$$

Special cases may arise in the calculation of n and m when M_1, M_2, and/or M_3 become zero.

If M_1, M_2, and M_3 are all zero, then the surface reduces to a plane with linear isoparametric curves. In this case, both m and n are set equal to one because

$$R_{11}(u, v) = 0, R_{22}(u, v) = 0, \text{ and } R_{12}(u, v) = 0.$$

If M_1 and M_3 are both zero, then the surface is flat along the isoparametric curves and both m and n are treated as equal. This accounts for linearity of isoparametric curves used in the surface/surface intersection problem.

If $M_1 = 0$, then the surface is flat in the u-direction and n is set equal to 1.

If $M_3 = 0$, then the surface is flat in the v direction and m is set equal to 1.

If M_1 and M_3 are both nonzero, then the calculated mixed partial derivative term is rolled proportionally in u and v steps based on M_1/M_3 (i.e., n/m is set to be M_1/M_3).

This method depends on the direct evaluation of second-order partial derivatives and their 2-norms. There are two problems associated with this method. First, the second-order derivatives may not exist for the surface, as in the case of C^1 surfaces. Secondly, even if the derivatives exist, it may be time consuming to compute these derivatives and the bounds on the 2-norms of these derivatives. No doubt, it is easy to implement the calculation of the derivatives in the case of polynomial surfaces. In particular, these derivatives for cubic parametric surfaces reduce to linear terms and it is trivial to calculate bounds on linear expressions.

Let K be defined as

$$K = (M_1/n^2 + 2M_2/mn + M_3/m^2)/8.$$

To calculate the bounding volume for a surface, first the term K is computed and then Max' and Min' are calculated as the maximum and minimum values of $R(p, q)$, $R(p + dp, q)$, $R(p, q + dq)$, and $R(p + dp, q + dq)$—four corners of a surface piece. The bounding volume of this surface piece is than obtained by using K as the expansion factor:

$$\text{Min}(i) = \text{Min}'(i) - K,$$
$$\text{Max}(i) = \text{Max}'(i) + K, \qquad \text{for } i = 1, 2, \text{ and } 3.$$

This technique guarantees that the entire surface piece lies inside the bounding volume. It is based on the ability to calculate the second-order derivatives, to calculate the bounds on them, and finally to calculate the maximum and minimum of the corner points. This method is very successful with polynomial surfaces, specifically with cubic polynomial surfaces. However, this technique is not practical at all for general parametric surfaces. The difficulty lies in the calculation of the second-order derivatives for general parametric surfaces where the derivatives may not exit. Even when the derivatives exist, it is not easy to compute the norms on these derivatives. This problem makes such a method very clumsy to use in real time applications. In general, these norms are not used in CAGD [11].

2.2. Details of Surface-Oriented Method

This method differs from the axis-oriented method, SURF81/SURF87, discussed in Section 2.1.1., because the bounding volumes are not oriented along the coordinate axes. Rather, they are oriented along the surface involving the positioning of the surface. Such bounding volumes are supposed to yield smaller geometric volumes enclosing the surfaces [6]. For computation-intensive applications it is not only desirable but also necessary to minimize the number of bounding volumes. Thus, it was assumed that, in general applications, the surface-oriented bounding volumes will be fewer in number than the axis-oriented bounding volumes.

A local coordinate system for the surface piece for the surface $R(u, v)$ is calculated on the parameter rectangle $[u_i, u_t] \times [v_i, v_t]$, where u_i, v_i are the initial values of the parameters and u_t, v_t are the terminal values of the parameters. Define a unit vector e_1 in terms of the position values along the $v = v_i$ or $v = v_t$ parametric curves, if possible. Otherwise, consider e_1 to be the unit vector along the positive direction of the x-axis of the coordinate system. Similarly, if possible, define a unit vector e_2 in terms of the position values along the $u = u_i$ or $u = u_t$ parametric curves. Otherwise, consider e_2 to be the unit vector along the positive direction of the y-axis of the coordinate system. Once two noncollinear vectors are determined, the third unit vector is determined by the relation

$$e_3 = (e_1 \times e_2)/|e_1 \times e_2|$$

where "$\times$" denotes the cross-product between the vectors.

Since e_1 and e_2 are not necessarily orthogonal, the unit vector e_2 is recalculated as

$$e_2 = (e_3 \times e_1)/|e_3 \times e_1|.$$

The resulting vectors e_1, e_2, and e_3 form a local orthonormal system. Relative to this orthonormal system, Min and Max are calculated as in Section 2.1.1. The surface-oriented bounding volume is defined in terms of three components:

1. Min is the anchor point.
2. The sides are oriented along the direction vectors of the local orthonormal system and they emanate from the anchor point.
3. The lengths of the sides of the bounding volume are determined by using the values of Min and Max computed above.

The intersection condition between two bounding volumes is determined after performing the following two steps:

1. Transform one of the bounding volumes in such a way that the anchor point coincides with the origin and the sides are oriented along the positive directions of x-, y-, and z-axis.

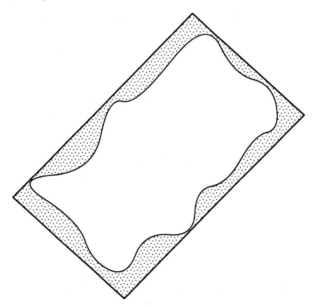

FIGURE 1.3.   Surface-Oriented Bounding Box

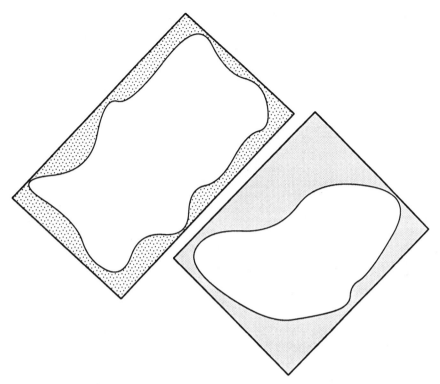

FIGURE 1.4.   Surface-Oriented Boxes Do Not Intersect, Surfaces Do Not Intersect

2.   Apply the same transformation to the second bounding volume. The inter-
     section condition is evaluated by performing the intersection of the edges of
     one box with the plane faces enclosing the other bounding volume and vice
     versa.

2.3.   Details of Convex Hull Method

This method differs from methods discussed in Sections 2.1.1, 2.1.2, and 2.2
because a convex hull is not necessarily a rectangular parallelepiped. Rather, a
convex hull is a systematic collection of planes wrapping around the surface such
that the enclosing volume is minimal. There is less deviation between the volume
of a convex hull and the geometric volume of a surface than with the previous
methods.

     The calculation of convex hulls is a function of control points of such spe-
cialized surfaces as B-spline and Bezier surfaces. The calculation of convex hulls

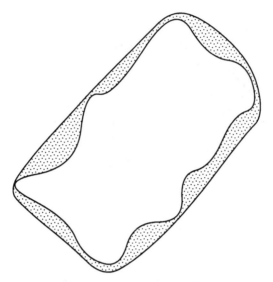

FIGURE 1.5.   Convex Hull

for general parametric surfaces is a hard nut to crack. The computation of the convex hulls and the intersection between the convex hulls for general parametric surfaces is more complex than it is for surface-oriented boxes. Literature is full of techniques for evaluating the convex hulls [12]. We will not delve into the details of convex hull calculations because it is not the way to approach bounding volumes for arbitrary parametric surfaces.

2.4.   Details of Ellipsoidal Method

This method is applicable to C^k, $k \geq 1$, surfaces only [8]. This procedure depends on the ability to calculate the first-order partial derivatives of the surface and the bounds on these derivatives. This technique differs from all the methods discussed in Sections 2.1.1, 2.1.2, 2.2, and 2.3, where the bounding volumes are enclosed by planar faces. Here the bounding volumes themselves are curved surfaces and they are ellipsoidal or spheroidal in nature.

The discussion of this method warrants the discussion of some terminology. Let the surface $R(u, v)$ be defined over the rectangle $[u_i, v_i] \times [u_t, v_t]$, where u_i, v_i are the initial values of the parameters, and u_t, v_t are the terminal values of the parameters u, v. The 2-norm of A is denoted by $\|A\|$ and is defined by the square root of the sum of the squares of the components of A. Let K be the Lipschitz constant for the surface $R(u, v)$, such that

$$\|R(u, v) - R(u_1, v_1)\| \leq K\|(u, v) - (u_1, v_1)\|.$$

Since

$$\|(u, v) - (u_1, v_1)\| \le (|u - u_1| + |v - v_1|)$$

the Lipschitz condition can be replaced by a simpler modified condition

$$\|R(u, v) - R(u_1, v_1)\| \le K(|u - u_1| + |v - v_1|).$$

This modified condition is used to calculate the ellipsoidal bounding volume with two foci at $R(u_i, v_i)$ and $R(u_t, v_t)$, and with the length of major axis as

$$L = K(|u_i - u_t| + |v_i - v_t|).$$

The resulting ellipsoid becomes

$$\|R(u, v) - R(u_i, v_i)\| + \|R(u, v) - R(u_t, v_t)\| = L.$$

The calculation [8] of the Lipschitz constant, K, is found for C^1 continuous surfaces as $K =$

$$\sup(\|R_1(u, v)\| + \|R_2(u, v)\|).$$

Here $R_1(u, v)$ is the first-order partial derivative vector with respect to u, and $R_2(u, v)$ is the first-order partial derivative vector with respect to v of $R(u, v)$; sup is the maximum value of the sum of the 2-norms of the first-order derivatives of $R(u, v)$ over the defining rectangle

$$[u_i, v_i] \times [u_t, v_t].$$

If $R(u_i, v_i) = R(u_t, v_t)$ then ellipsoids reduce to spheroids. For brevity, the ellipsoid with foci P_1, P_2, and major axis length L_P is denoted by $E_P = (P_1, P_2, L_P)$.

The ellipsoidal technique is feasible for surfaces such as bicubics or low-order polynomials because local maximums of the parametric derivatives are easy to evaluate. If the surface is piecewise continuously differentiable, the maximum of each piece may be used. The application must know, in advance, the number of pieces used in the definition. This method at least guarantees that the entire surface will lie inside the bounding volume. Many applications, such as surface/surface intersection, use the bounding volumes for the culling process and thus do not require this guarantee of numerical accuracy. In such cases, sampled positional data may be used for calculating the derivatives [13].

The intersection condition can be determined by using the foci and Lipschitz constant for one bounding volume and the sampled positional data of the second bounding volume. That is, two ellipsoidal bounding volumes,

$$E_P = (P_1, P_2, L_P) \text{ and}$$
$$E_Q = (Q_1, Q_2, L_Q),$$

intersect if there exists a point P on E_P such that

$$\|P - Q_1\| + \|P - Q_2\| \le L_Q \text{ or}$$

there exists a point Q on E_Q, such that

$$\|Q - P_1\| + \|Q - P_2\| \le L_P.$$

This method is simpler than the surface-oriented technique and more complex than the axis-oriented method for the calculation and intersection of bounding volumes. However, it depends on the availability of calculations for the first-order partial derivatives and the norms on them. Hence, this method is not applicable to arbitrary parametric surfaces. Since axis-oriented and surface-oriented methods both use positional data, they are applicable to a more general class of parametric surfaces. These methods will be the prime candidates for run-time analysis in this discussion.

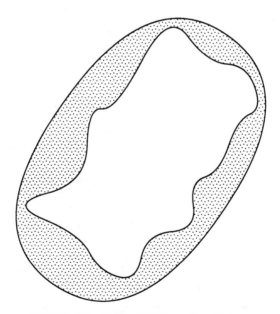

FIGURE 1.6. Ellipsoidal Bounding Volume

2.5.   Details of Spheroidal Method

This method is applicable to all C^k, $k \geq 1$, surfaces [9, 10]. Like the ellipsoidal procedure discussed in Section 2.4, this procedure also depends on the ability to calculate the first-order partial derivatives of the surface and the bounds on these derivatives. However, this technique differs from the ellipsoidal method in several ways.

Let the surface $R(u, v)$ be defined over the rectangle $[u_i, v_i] \times [u_t, v_t]$ where u_i, v_i are the initial values of the parameters, and u_t, v_t are the terminal values of the parameters u, v. An ellipsoid with two foci at $R(u_i, v_i)$ and $R(u_t, v_t)$, and with the length, L_{old} of major axis is defined in Section 2.4. as:

$$\|R(u, v) - R(u_i, v_i)\| + \|R(u, v) - R(u_t, v_t)\| = L_{\text{old}}$$

where

$$L_{\text{old}} = K_{\text{old}}(|u_i - u_t| + |v_i - v_t|).$$

Note that the term "old" is used in the context of values obtained in Section 2.4, and the term "new" is used in context of the values calculated using the techniques of this section. The value of the Lipschitz constant [8] is determined for C^1 continuous surfaces as

$$K_{\text{old}} = \sup (\|R_1(u, v)\| + \|R_2(u, v)\|).$$

Here $R_1(u, v)$ is the first-order partial derivative vector with respect to u, and $R_2(u, v)$ is the first-order partial derivative with respect to v of $R(u, v)$; sup is the maximum value of the sum of the 2-norms of the first-order derivatives of vector $R(u, v)$ over the defining rectangle $[u_i, v_i] \times [u_t, v_t]$.

The ellipsoid with foci P_1, P_2, and major axis length as L_P is denoted by $E_P = (P_1, P_2, L_P)$. It may be observed that the "old" analysis uses a gross estimate of Lipschitz constant. This L_{old} does not take advantage of the location of the foci (i.e., the same L_{old} can be used with foci at:

$R(u_i, v_i)$, and $R(u_t, v_t)$, or
$R(u_i, v_t)$, and $R(u_t, v_i)$, or
$R(u_i, v_i)$, and $R(u_i, v_t)$, or
$R(u_i, v_i)$, and $R(u_t, v_i))$.

Even a spheroid with radius L_{old} and center at any of the above points also represents a bounding volume. This general purpose L_{old} is not then a good estimate for calculating the bounding volumes.

The evaluation of intersection condition is another source of inefficiency embedded in the ellipsoidal bounding volumes. The intersection condition is determined by using the foci and the Lipschitz constant for one bounding volume and the sampled positional data on the second bounding volume. That is, two ellipsoidal bounding volumes,

$$E_P = (P_1, P_2, L_P) \text{ and}$$
$$E_Q = (Q_1, Q_2, L_Q),$$

intersect if there exists a point P (sampled or otherwise) on ellipsoid E_P, such that

$$\|P - Q_1\| + \|P - Q_2\| \leq L_Q \text{ or}$$

there exists a point Q on ellipsoid E_Q, such that

$$\|Q - P_1\| + \|Q - P_2\| \leq L_P.$$

This analysis indicates that the ellipsoidal method is not the best choice for C^1 surfaces. The ellipsoidal method capitalizes on the 1-norm, which is, in general, larger than the 2-norm, which, in turn, is larger than the 0-norm.

Because

$$\|R(u, v)\|_0 \leq \|R(u, v)\|_2 \leq \|R(u, v)\|_1,$$

the new technique, developed here, utilizes a hybrid scheme involving the 0-norm and 2-norm. Define

$$m(u, v, i) = \max(|R_1(u, v, i)|, |R_2(u, v, i)|)$$

for $i = 1, 2, 3$.

Here $R_1(u, v, i)$ is the first-order partial derivative of the i-th coordinate of $R(u, v)$ with respect to u, and $R_2(u, v, j)$ is the first-order partial derivative of the j-th coordinate of $R(u, v)$ with respect to v for $1 \leq i, j \leq 3$.

Now

$$\|R(u, v) - R(p, q)\| = \|(u - p)R_1(\underline{u}, \underline{v}) + (v - q)R_2(\underline{u}, \underline{v})\|$$

where $\underline{u}$ lies between u and p, and $\underline{v}$ lies between v and q. It follows that

$$\begin{aligned} \|R(u, v) - R(p, q)\|^2 \leq [&((u - p)R_1(\underline{u}, \underline{v}, 1) + (v - q)R_2(\underline{u}, \underline{v}, 1))^2 + \\ &((u - p)R_1(\underline{u}, \underline{v}, 2) + (v - q)R_2(\underline{u}, \underline{v}, 2))^2 + \\ &((u - p)R_1(\underline{u}, \underline{v}, 3) + (v - q)R_2(\underline{u}, \underline{v}, 3))^2] \end{aligned}$$

$$\leq [(|u - p|m(\underline{u}, \underline{v}, 1) + |v - q|m(\underline{u}, \underline{v}, 1))^2 +$$
$$(|u - p|m(\underline{u}, \underline{v}, 2) + |v - q|m(\underline{u}, \underline{v}, 2))^2 +$$
$$(|u - p|m(\underline{u}, \underline{v}, 3) + |v - q|m(\underline{u}, \underline{v}, 3))^2]$$

$$\leq [(|u - p| + |v - q|)^2 m(\underline{u}, \underline{v}, 1)^2 +$$
$$(|u - p| + |v - q|)^2 m(\underline{u}, \underline{v}, 2)^2 +$$
$$(|u - p| + |v - q|)^2 m(\underline{u}, \underline{v}, 3)^2]$$

$$\leq [(|u - p| + |v - q|)^2$$
$$(m(\underline{u}, \underline{v}, 1)^2 + m(\underline{u}, \underline{v}, 2)^2 + m(\underline{u}, \underline{v}, 3)^2)]$$
$$\leq K_{\text{new}}^2 [(|u - p| + |v - q|)^2].$$

Here

$$K_{\text{new}} = \sup \|((m(u, v, 1), m(u, v, 2), m(u, v, 3))\|),$$

such that sup is the maximum value of 2-norm of vector $(m(u, v, 1), m(u, v, 2),$ $m(u, v, 3))$ over the surfaces defining rectangle $[u_i, v_i] \times [u_t, v_t]$.

It is clear that K_{new} is much smaller than K_{old}, which means that the old value K_{old} can be replaced by K_{new}.

The method with modified norm, again, guarantees that the entire surface lies inside the new ellipsoidal bounding volume with $L_{\text{new}} = K_{\text{new}}(|u_i - u_t| + |v_i - v_t|)$. Since it is easier to use a spheroid for intersection detection than an ellipsoid, a spheroid with center at

$$(R(u_i, v_i) + R(u_t, v_t))/2, \text{ or}$$
$$(R(u_i, v_t) + R(u_t, v_i))/2 \qquad \text{or even with}$$
$$R(u_i, v_i) + R(u_t, v_t) +$$
$$R(u_i, v_t) + R(u_t, v_i))/4$$

and with radius equal to $L_{\text{new}}/2$ can serve as a bounding volume. The calculation of this spheroidal bounding volume is performed as follows:

$$\|R(u, v) - (R(u_i, v_i) + R(u_t, v_t) + R(u_i, v_t) + R(u_t, v_i))/4\|$$
$$\leq (1/4) (\|R(u, v) - R(u_i, v_i)\|$$
$$+ \|R(u, v) - R(u_t, v_t)\|$$
$$+ \|R(u, v) - R(u_i, v_t)\|$$
$$+ \|R(u, v) - R(u_t, v_i)\|)$$

$$\leq (1/4)(K_{\text{new}}(|u_i - u_t| + |v_i - v_t|)$$
$$+ K_{\text{new}}(|u_i - u_t| + |v_t - v_i|))$$
$$\leq (1/2)K_{\text{new}}(|u_i - u_t| + |v_i - v_t|)$$
$$\leq (1/2)L_{\text{new}}.$$

This approach to spheroids depends implicitly on first computing the ellipsoidal information. Also, the center of the spheroid does not lie on the surface. We consider a method for calculating the spheroids that is independent of the ellipsoidal information and the center of the spheroid lies on the surface. To illustrate the derivation of spheroidal method, the following mechanism is used. Let C be a point on the surface of the spheroid corresponding to the parametric midpoint

$$((u_i + u_t)/2, \quad (v_i + v_t)/2)$$

of the parametric space defining the surface $R(u, v)$. Using the calculations for K_{new}, it follows that

$$\begin{aligned}
\|R(u, v) - C\| &= \|(u - (u_i + u_t)/2)R_1(\underline{u}, \underline{v}) \\
&\quad + (v - (v_i + v_t)/2)R_2(\underline{u}, \underline{v})\| \\
&\leq K_{new}(|u_i - u_t| + |v_i - v_t|)/2
\end{aligned}$$

where $(\underline{u}, \underline{v})$ is a point in the parameter space of the surface, such that $\underline{u}$ lies between u and p, and $\underline{v}$ lies between v and q.

On the average, a spheroid with radius L_{new}, where

$$L_{new} = K_{new}\|(u_i - u_t, v_i - v_t)\|/2,$$

is 50 percent smaller than the "old" ellipsoid. This results in a significant increase in speed for calculating spheroidal bounding volumes over ellipsoidal bounding volumes.

As noted, the determination of the intersection condition between ellipsoidal bounding volumes is not easy either. However, it is very simple in the case of spheroidal bounding volumes because there is no need to consider the positional data. Let $S_1 = (C_1, r_1)$ be a spheroid with center C_1 and radius r_1, and let $S_2 = (C_2, r_2)$ be a spheroid with center C_2 and radius r_2. The two spheroids intersect if

$$\|C_1 - C_2\| \leq (r_1 + r_2).$$

In the spheroidal method, the only calculation needed is the distance between the centers. The ellipsoidal method uses the defining property of one ellipsoid along with the positional data on the other ellipsoid.

Finally, even though this method depends on the availability of the first-order derivatives and the norms on them, it is simpler than the surface-oriented technique. It is more complicated than the axis-oriented method for the calculation and intersection of bounding volumes. Also, this method is not applicable to arbitrary parametric surfaces. Since axis-oriented and surface-oriented methods both use positional data, they are applicable to a more general class of parametric

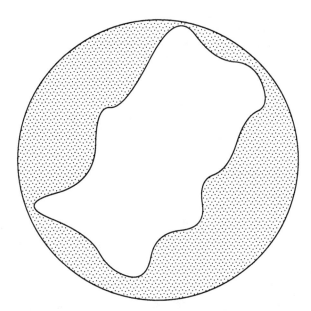

FIGURE 1.7.   Spheroidal Bounding Volume

surfaces. These two methods are the prime candidates for run-time analysis in this discussion.

3.   STORAGE REQUIREMENTS

Section 2 discusses five methods for computing bounding volumes for parametric surfaces. Computing volumes by means of axis-oriented (SURF87) and surface-oriented (SURF84) methods is easier than the convex hull, ellipsoidal, and spheroidal methods. In SURF84 and SURF87, the computations in source code are distributed over two procedures with different names used in different implementations. However, one of the procedures calculates the bounding volumes and the other procedure calculates the intersection between them. The methods for implementations SURF81 and SURF87 are identical as far as the calculation and intersection of bounding volumes are concerned, but they differ essentially in how these calculations are implemented. In SURF81, the position values are evaluated by calls to the parametric evaluators at the time of bounding volume calculation. However, in SURF87 the evaluator calls are stripped of the routines and the position values are precalculated—they are simply used here. The main difference in the structure of bounding volume calculation, is depicted between

TABLE 1.1. Storage Requirements

Routine	SURF81	SURF84	SURF87
Calcbox	X	319	X
Boxint	X	296	X
Compbox	58	X	54
Box	58	X	48
TOTAL LINES	116	615	102
TOTAL ROUTINES	2	2	2
LINES/ROUTINES	58	318	51

SURF84 and SURF87. Both have the identical input, but the procedures for the calculation and intersection of bounding volumes are absolutely different. For each of the two methods, Table 1.1 describes: (a) the routines used to calculate the bounding volumes for surface pieces, (b) the routines used to calculate the intersection between the bounding volumes for surface pieces, (c) the length of each routine, (d) total lines of source code used, and (e) the average number of source code lines used per routine. In Table 1.1, 'X' is used to indicate that a particular routine is not used in the method named in the column heading.

4. SAMPLE TEST CASES

For analyzing the execution time of these methods, we considered a wide variety of samples of bounding volumes. The data for the calculation and intersection of bounding volumes is drawn from various surfaces—sculptured, cylindrical, conical, spheroidal, toroidal, loft, and planar surfaces. By using the bounding volumes, these surfaces are selectively subdivided into smaller surface pieces with surface convexity and boundary curvature within ten degrees in each method. Table 1.2 shows the numbers of bounding volumes, that actually occurred in the above subdivision process, used in the three implementations SURF81, SURF84, and SURF87 for timing analysis of each model (that is, surface pair).

5. TIME COMPLEXITY

Since there are no known analytic performance results of these methods, we must rely on the empirical evidence that comes from extensive simulation as presented in Tables 1.3 through 1.10. Computing volumes by means of axis-oriented and surface-oriented methods is easier than the convex hull, ellipsoidal, and spheroidal methods. Here comparison is made between these two simpler methods—

TABLE 1.2.   Numbers of Bounding Volumes Used

Model	SURF81	SURF84	SURF87
CONCON	112	178	120
CONCYL	114	210	119
CYLPAIRP	92	176	94
EPLERR	27	33	31
FALSEJOI	17	25	19
KGKSIMPL	2	2	1
LOFT57	6	6	6
PLCONE	57	91	63
PLCYL	31	49	35
PLPL	2	2	1
PLSPH	125	257	121
PLSPHP	125	257	121
PLTORUS	129	225	93
SPHCON	150	296	162
SPHCYL	150	296	162

TABLE 1.3.   Worst Case CPU Times in Microseconds for Bounding Volume Calculation

Model	SURF81	SURF84	SURF87
CONCON	3854	885	729
CONCYL	3906	1042	729
CYLPAIRP	3906	833	781
EPLERR	3854	833	729
FALSEJOI	3906	781	729
KGKSIMPL	3854	729	729
LOFT57	3854	885	677
PLCONE	3854	937	729
PLCYL	3854	833	729
PLPL	3902	781	729
PLSPH	3906	937	729
PLSPHP	3854	937	729
PLTORUS	3854	937	729
SPHCON	3906	885	729
SPHCYL	3906	885	729
Overall average	3878	875	729
NORMALIZED	5.32	1.20	1.00

SURF84 and SURF87. The algorithms were implemented on an IBM 4381 series computer. Computation times vary from system to system, and even on the same system from hour to hour depending on the amount of memory load. These tests were made so that each method was tested on the same model and during the same memory load time. For the sake of determining the relative performance of the algorithms, we normalized the time requirements with respect to the least times.

5.1.   Bounding Volume Calculation Times

For calculating bounding volumes for single surface pieces, from Table 1.3 for the worst case execution times, it follows that SURF87 is 20 percent faster than SURF84; and SURF87 is 432 percent faster than SURF81. From Table 1.4 for the best case execution times, it follows that SURF87 is 74 percent faster than SURF84; and SURF87 is 787 percent faster than SURF81. The comparison by overall times as in Table 1.5 is more realistic than the times in Table 1.3 and Table 1.4, because a single best case or a single worst case example can bias the results of the entire set of examples. From Table 1.5 for the average execution times, it follows that SURF87 is 73 percent faster than SURF84; and SURF87 is 718 percent faster than SURF81.

It is a long-held belief that in spite of the computational complexity of the

TABLE 1.4.   Best Case CPU Times in Microseconds for Bounding Volume Calculation

Model	SURF81	SURF84	SURF87
CONCON	3750	833	365
CONCYL	3750	729	365
CYLPAIRP	3750	729	411
EPLERR	3750	729	365
FALSEJOI	3802	729	417
KGKSIMPL	3802	729	729
LOFT57	3802	833	417
PLCONE	3750	729	365
PLCYL	3750	729	365
PLPL	3750	729	729
PLSPH	3750	677	365
PLSPHP	3750	729	365
PLTORUS	3750	729	365
SPHCON	3750	729	365
SPHCYL	3750	729	365
Overall average	3760	739	424
NORMALIZED	8.87	1.74	1.00

TABLE 1.5.   Average Case CPU Times in Microseconds for Bounding Volume Calculation

Model	SURF81	SURF84	SURF87
CONCON	3810	836	425
CONCYL	3824	805	431
CYLPAIRP	3804	782	445
EPLERR	3791	778	423
FALSEJOI	3842	777	433
KGKSIMPL	3828	729	729
LOFT57	3828	859	469
PLCONE	3823	833	411
PLCYL	3795	780	421
PLPL	3776	755	729
PLSPH	3797	841	408
PLSPHP	3798	842	397
PLTORUS	3788	838	405
SPHCON	3797	815	425
SPHCYL	3795	813	424
Overall average	3806	806	465
NORMALIZED	8.18	1.73	1.00

bounding volumes, the surface-oriented method, SURF84, is superior to axis-oriented method SURF81/SURF87 in terms of overall performance. Unfortunately, this is not true. The proof of the pudding lies in the eating thereof. For the total times needed to calculate the bounding volumes for a surface pair intersection, it follows from Table 1.6 that SURF87 is 264 percent faster than SURF84.

5.2.   Bounding Volume Intersection Times

For intersection between the bounding volumes for single surface pairs, from Table 1.7 for the worst case execution times, it follows that SURF87 is 2322 percent faster than SURF84; and SURF87 is 129 percent faster than SURF81. From Table 1.8 for the best case execution times, it follows that SURF87 is 738 percent faster than SURF84. The comparison by overall times as in Table 1.9 is more realistic than the times in Table 1.7 and Table 1.8 because a single best case or a single worst case example can bias the results of the entire set of examples. From Table 1.9 for the average execution times, it follows that SURF87 is 2066 percent faster than SURF84; and SURF87 is 51 percent faster than SURF81.

It is a long-held and much debated belief that in spite of the intersection complexity of the surface-oriented bounding volumes, the surface-oriented method, SURF84, is superior to axis-oriented method, SURF81/SURF87, in terms of

TABLE 1.6.   Total CPU Times in Microseconds
for Bounding Volume Calculations

Model	SURF81	SURF84	SURF87
CONCON	426667	148802	51042
CONCYL	435937	168958	51250
CYLPAIRP	350000	137552	41823
EPLERR	102344	25677	4479
FALSEJOI	65312	19423	8229
KGKSIMPL	7656	1458	729
LOFT57	22969	5156	2344
PLCONE	217917	75833	25885
PLCYL	117656	38229	14740
PLPL	7552	1510	729
PLSPH	474635	216250	49323
PLSPHP	474739	216406	48073
PLTORUS	488698	188490	37656
SPHCON	569531	241146	68854
SPHCYL	569323	240677	68646
Overall average	288729	115838	31589
NORMALIZED	9.14	3.66	1.00

TABLE 1.7.   Worst Case CPU Times in Microseconds
for Bounding Volume Intersection

Model	SURF81	SURF84	SURF87
CONCON	312	5000	156
CONCYL	260	4948	208
CYLPAIRP	260	4948	156
EPLERR	260	1927	104
FALSEJOI	312	1875	104
KGKSIMPL	260	521	104
LOFT57	312	4687	104
PLCONE	260	729	104
PLCYL	260	1875	104
PLPL	260	521	104
PLSPH	260	2031	104
PLSPHP	260	1927	104
PLTORUS	208	1927	104
SPHCON	312	5000	104
SPHCYL	260	4948	104
Overall average	270	2858	118
NORMALIZED	2.29	24.22	1.00

TABLE 1.8. Best Case CPU Times in Microseconds
for Bounding Volume Intersection

Model	SURF81	SURF84	SURF87
CONCON	104	729	104
CONCYL	104	729	104
CYLPAIRP	104	990	104
EPLERR	104	1562	104
FALSEJOI	104	521	104
KGKSIMPL	104	521	104
LOFT57	104	1875	104
PLCONE	104	677	104
PLCYL	104	677	104
PLPL	104	521	104
PLSPH	104	1510	104
PLSPHP	104	677	104
PLTORUS	104	625	104
SPHCON	104	729	104
SPHCYL	104	729	104
Overall average	104	872	104
NORMALIZED	1.00	8.38	1.00

TABLE 1.9. Average Case CPU Times in Microseconds
for Bounding Volume Intersection

Model	SURF81	SURF84	SURF87
CONCON	110	3295	104
CONCYL	113	3881	105
CYLPAIRP	121	3938	104
EPLERR	170	1701	104
FALSEJOI	177	1415	104
KGKSIMPL	182	1521	104
LOFT57	193	2990	104
PLCONE	181	678	104
PLCYL	177	924	104
PLPL	182	521	104
PLSPH	171	1968	104
PLSPHP	183	1296	104
PLTORUS	168	1421	104
SPHCON	117	4123	104
SPHCYL	116	4120	104
Overall average	157	2253	104
NORMALIZED	1.51	21.66	1.00

TABLE 1.10.   Total CPU Times in Microseconds
for Bounding Volume Intersection

Model	SURF81	SURF84	SURF87
CONCON	176094	3795980	177240
CONCYL	120625	3858541	115208
CYLPAIRP	50312	2898175	48125
EPLERR	5104	54427	4479
FALSEJOI	3177	33958	2604
KGKSIMPL	365	521	417
LOFT57	1354	14948	937
PLCONE	13021	60990	9479
PLCYL	6354	44375	5104
PLPL	365	521	417
PLSPH	23906	434792	15521
PLSPHP	25573	331719	15521
PLTORUS	23229	318333	11771
SPHCON	97552	798020	98906
SPHCYL	97083	724582	98854
Overall average	42941	891325	40306
NORMALIZED	1.07	22.11	1.00

overall performance. It is shown here by Table 1.10 that this is a false assumption. For total intersection times, for all the bounding volumes needed for the intersection of a surface pair, it follows from Table 1.10 that SURF87 is 2111 percent faster than SURF84.

6.   CONCLUSIONS AND FURTHER STUDY

Here we identify what has been accomplished, and what needs to be analyzed further. The benchmark study for the calculation and intersection between bounding volumes as used in CAD/CAM applications has been discussed in detail. It has been a wishful hypothesis that in spite of the computational complexity of the bounding volumes, the surface oriented method, SURF84, is superior to axis-oriented method, SURF81/SURF87, in terms of overall performance. At the time of designing and implementing SURF84, it was suggested that surface-oriented boxes may not be worth the effort, but, in turn, it was believed that the long-term benefits of this method would far outweigh the cost for generating surface oriented volumes. However, this turns out to be false perception. For average times needed to calculate a bounding volume, it is clear from Table 1.5 that SURF87 is 73 percent faster than SURF84. For average times needed to detect the intersection condition between two bounding volumes, it has been determined from Table 1.9

that SURF87 is 2066 percent faster than SURF84. For the total times needed to calculate and intersect the bounding volumes for a surface pair intersection, it has been determined from Table 1.6 that SURF87 is 264 percent faster than SURF84 for calculating bounding volumes, and from Table 1.10 that SURF87 is 2111 percent faster than SURF84 for intersecting the bounding volumes. Thus, it is determined that SURF87 is significantly faster than any other method as far as the calculation of bounding volumes and intersection between the bounding volumes is concerned; whether it is used on individual surface pieces or an entire set of surface pairs encountered in an application.

Analytically, it is apparent that it is easier to compute bounding volumes by C^1 ellipsoidal method than by C^2 axis-oriented method, whereas it is easier to detect the intersection condition between two bounding volumes by the C^2 axis-oriented method than by the C^1 ellipsoidal method. If one is working with sufficiently smooth surfaces, it is worth gathering empirical data to determine the suitability of these methods for production modules. It is desirable to compare the performance of C^1 spheroids with the C^1 ellipsoids. It is anticipated that performance comparison will be on the same scale as that of axis-oriented and surface-oriented bounding volumes. But, it has been determined that the C^0 axis-oriented method is simpler and more versatile than any other method for both calculation and intersection of bounding volumes.

REFERENCES

1. T.G. Melson, "Surface/Surface Intersection," *MCAIR Report 78-011*, McDonnell Douglas Aircraft Company, St. Louis, MO, 1978, pp. 1–14.
2. C.L. Sabharwal, "A Divide-and-Conquer Method for Curve Intersection between Two C^1 Parametric Surfaces," *Advanced CAD/CAM Technology Technical Report 25-009*, McDonnell Douglas, St. Louis, MO, 1981a, pp. 1–66.
3. C.L. Sabharwal, "A Divide-and-Conquer Method for Curve Intersection Between Two Parametric Surfaces," *Advanced CAD/CAM Technology Research Technical Note 25-010* McDonnell Douglas, St. Louis, MO, 1981b, pp. 1–172.
4. M. Berger, *Computer Graphics with Pascal*, Benjamin-Cummings, CA, 1986.
5. M. Segal and C.H. Sequin, "Partitioning of Polyhedral Objects into Disjoint Parts," *Computer Graphics and Applications*, Vol. 8, 1988, pp. 53–67.
6. E.G. Houghton, R.F. Emnett, J.D. Factor, and C.L. Sabharwal, "Implementation of a Divide-and-Conquer Method for Intersection of Parametric Surfaces," *Computer Aided Geometric Design*, Vol. 2, 1985, pp. 173–183.
7. R.N. Goldman and T.D. Derose, "Recursive Subdivision without the Convex Hull Property," *Computer Aided Geometric Design*, Vol. 3, 1986, pp. 247–265.
8. B.V. Herzen and A.H. Barr, "Accurate Triangulation of Deformed, Intersection Surfaces," *Computer Graphics*, Vol. 21, 1987, pp. 103–110.
9. C.L. Sabharwal and T.G. Melson, "Implementation of Cross Intersection Between Triangular Surfaces," *Proceedings of ACM Annual Computer Science Conference '88*, 1988a.

10. C.L. Sabharwal and T.G. Melson, "Survey of Bounding Volumes for Parametric Surfaces," *MCAIR Report B0820*, McDonnell Douglas, St. Louis, MO, 1988b.
11. D. Filip, R. Magedson, and R. Markot, "Surface Algorithms using Bounds on Derivatives," *Computer Aided Geometric Design*, Vol. *3*, 1986, pp. 295–311.
12. R.H. Bartels, J.C. Beatty, and B.A. Barsky, *An Introduction to Splines for use in Computer Graphics and Geometric Modeling*, Morgan Kauffman, CA, 1987.
13. H. Akima, "On Estimating Partial Derivatives for Bivariate Interpolation of Scattered Data," *Rockey Mountain Journal of Mathematics*, Vol. *14*, 1984, pp. 41–52.

2

Motion Learning in Robotized Mechanical Systems*

Giuseppe Casalino
Michele Aicardi
Department of Communications,
Computers and Systems Science
(DIST)-University of Genoa
Genova
Italy

1. INTRODUCTION

Roughly speaking, a learning system is intended to be a structure capable of improving its performances on the basis of the experience acquired during the execution of the same or similar tasks.

In the robotic-control field, this gross definition has been recently specialized to the problem of eventually allowing a manipulator to exactly execute a planned motion, even in the presence of (possibly very high) uncertainties in the mathematical model of the manipulator itself. Methodologies dealing with the problem belong to the class of the so-called "Iterative Learning Techniques" recently and almost independently introduced by many researchers [1, 2, 3, 4, 5, 6, 7, 8, 9, 10]. Learning control techniques are structured in such a way that trajectory errors occurring during the execution of a motion trial are used to reduce the errors that will occur during the next trial, until the eventual convergence accomplishes a perfect execution.

Like another class of control techniques, the so-called "Adaptive Control Techniques" (see, for instance, [11] for a good survey on the subject, or [12] as an introductory text), Iterative Learning Techniques are important because of the natural need to provide robots (possibly programmed to execute the same operations for many hours) with facilities for improving their performances through practice.

However, if compared with the adaptive ones, the learning technique exhibits

---

* This work has been partially supported by CNR special project on Robotics and European Project FIRST on Basic Research Initiatives.

a substantial difference. Actually, in the former, robot model parameters (and possibly regulator parameters) are recursively identified and updated, and in the latter, what is actually learned is the control signal allowing the required motion. (This is almost achieved without requiring the knowledge of a parametric model of the robot.)

Though adaptive control techniques are, at least in principle, more general then the learning ones, they suffer from the drawback of being applicable only when the system is not too difficult to be parametrically modeled, and the resulting model is not too complex to be handled in real time. In fact, it must be noted that even for simple robotic structures (two or three rotational degrees of freedom) adaptive control techniques tend to produce control structures generally characterized by a high degree of complexity.

On the contrary, as will be clarified in this chapter, learning techniques always can be implemented very simply (independently from the complexity of the robotic structure), they require almost no knowledge about the system model, and, nevertheless, they can guarantee very high accuracy of motion. However, an obvious disadvantage of such techniques is that they require the presence of repetitive situations (which are quite common in the industrial field) and a new set of trials must be performed with any new planned trajectory.

As a consequence of these considerations, it then appears that such previously mentioned items as modeling and control complexity, and required accuracy for important tasks and their repeatibility, should constitute the correct decision base for a possible choice between the two classes of techniques. However, pragmatically enough it is important that further research studies should also be directed toward an exploration of the possibilities of integration between the two methodologies. Therefore, Section 5 of this chapter presents a specific parametric identification algorithm that is suitable for use as a complementary tool within a learning environment. This might be considered as a first step toward an effective integration of the two methodologies.

As to the learning techniques themselves, before beginning a detailed discussion it must first be observed that their successful application must rely on some theoretical guidelines concerning both the following aspects: (1) condition for the uniform boundedness of the trajectory errors at any time of the duration interval of the trial and for any trial, and (2) conditions for the eventual convergence to zero of the errors at any instant of the duration time, for increasing number of trials.

Actually, of these two items, the first seems to be the most delicate when referring to robotic manipulators. It is the authors' opinion that for the time being it has not been considered with the required accuracy, at least within the current literature, which has been much more concerned with the second aspect. To clarify the point, note, for instance, that Arimoto et al. [3] provide results concerning both items when referring to learning techniques applied to linear systems, but assume the uniform boundedness of the errors (Aspect (1)) when extending the methodology to nonlinear robotic systems. Other authors consider small deviations with respect to the required trajectory and refer the application of the learning processes to linear approximations of the system, thus implicitly avoiding the problem in the general case. Hauser [5], instead, gives valid proof for an even wider class of nonlinear systems, whose equations satisfy some global Lipschitz conditions. However, such global conditions cannot, in general, be completely guaranteed when referring to robots or complex nonlinear mechanical systems (see Section 4.4.1 for major details).

On the other hand, Kawato et al. [10] present interesting considerations based on functional analysis concepts. Unfortunately, due to some lack of completeness in their proof, the provided conditions do not seem to be, by themselves, completely sufficient for guaranteeing boundedness and convergence within arbitrarily assigned duration times of the trials. Finally, some of the authors [9] provide conditions for the fulfillment of both Aspects (1) and (2). Their results, however, hold only under restrictive assumptions about the presence of real-time acceleration feedbacks acting on the robotic structure. In fact, since perfect acceleration feedbacks cannot be assumed for mechanical systems (otherwise they would introduce physically unrealizable "algebraic loops"), it immediately follows that the results of [9] actually refer to ideal situations that can only be approximated by suitable choices of the operating conditions. (See [9] for further details.)

As a consequence of these considerations, one of the objectives of this chapter is to present a global stability and convergence theory for learning procedures applied to fully nonlinear robotic systems without assuming the presence of any real-time acceleration feedback.

To this aim, and after a discussion about the structure and rationale underlying the learning procedures performed in Section 2, stability and convergence conditions will be obtained as a result of the developments presented in Sections 3 and 4.

Such results, mainly expressed in terms of the accuracy required of the a priori estimate of the inertia matrix of the robot, will also take into account the influence exerted by the position and velocity feedbacks acting on the system, as well as the duration time assigned to each trial.

Subsequently, in Section 5, we shall propose a simple specific algorithm for the identification of the inertia matrix parameters that are useful for the achievement of the required accuracy if it cannot be a priori guaranteed. Some preliminary connections with the class of adaptive control techniques will be discussed in this section.

Finally, a simulation example will be presented in Section 6. In Section 7, some remarks concerning open problems and possible future research directions will conclude the chapter.

2.   THE LEARNING PROCEDURE

Consider the set of differential equations representing the dynamic behavior of a robotized mechanical system, relating vector $q \in \Re^n$ of its (generalized) coordinates with vector $M \in \Re^n$ of the corresponding (generalized) torques, and having the general form

$$A(q)\ddot{q} + B(q, \dot{q})\dot{q} + C(q) = M \qquad (1)$$

where matrices $A(q)$ (positive definite), $B(q, \dot{q})$, and vector $C(q)$, keep into account the effects due to inertias, centrifugal/Coriolis (and possibly viscous) terms and gravitational actions, respectively.

As it is commonly recognized, provided a perfect knowledge of the mathematical model (1) is available (and easy to be handled in real time), a very good control strategy, suitable to be used for tracking any reference trajectory $q^*(t)$, is represented by the so-called "computed torque control method," which requires the application of torque vectors having the exact form

$$M = A(q)[\ddot{q}^* - L\delta\dot{q} - P\delta q] + B(q, \dot{q})\dot{q} + C(q) \qquad (2)$$

where P and L represent constant feedback matrices, that take into account the position and velocity error vectors, $\delta q \triangleq q - q^*$ and $\delta\dot{q}$, respectively.

As it can be easily verified, the application of control law (2) would introduce exact compensation of all nonlinearities, thus, leading to the following linear closed-loop error dynamic equation:

$$\delta\ddot{q} + L\delta\dot{q} + P\delta q = 0. \qquad (3)$$

Such closed loop behavior would, in turn, guarantee a perfect tracking whenever null initial error conditions occur (i.e., $\delta q(0) = \delta\dot{q}(0) = 0$), or at least an asymptotic perfect tracking (in case of nonnull initial error conditions), whose convergence rate could be imposed on the basis of an appropriate choice of the gain matrices L and P, namely by assigning the desired zeroes to the characteristic equation

$$det[s^2 + Ls + P] = 0. \qquad (4)$$

However, it is a matter of fact that situations where the system model is unknown or at least partially known, or even too complex to be real time handled, actually represent the rule more than the exception. In all these cases, (as also suggested by Craig [12]) we may, however, still refer to the control law (2) as applied in its "approximated form," that is,

$$M = \hat{A}(q)(\ddot{q}^* - L\delta\dot{q} - P\delta q) + \hat{B}(q, \dot{q})\dot{q} + \hat{C}(q) \tag{5}$$

where $\hat{A}(q)$, $\hat{B}(q, \dot{q})$, and $\hat{C}(q)$ represent "estimates" of the corresponding terms, at the extent which is allowed for the particularly considered application.

Note that in (5) it is also possible to consider very rough estimates, such as, for instance, constant matrices and vectors, and possibly a null matrix for $\hat{B}(q, \dot{q})$. In this way, (5) can be seen as a very general controller structure, which also results to be comprehensive of the usual proportional-derivative linear regulators, commonly used in almost all industrial robotic applications.

Obviously, a robot controller belonging to class (5) can only allow an approximated tracking of the reference trajectory $q^*(t)$ (even in the presence of null initial-error conditions), whose accuracy clearly depends not only on the "amount" of linear feedback (L, P) and on the quality of the estimates, but also on the characteristics of the assigned reference trajectory (see, for instance, Craig [12] for more detailed considerations on this subject).

In this chapter, we shall refer to robotic structures equipped with regulators of the form (5), but slightly modified by the assumed presence of a manipulable external signal u, as indicated in the following regulator espression

$$M = \hat{A}(q)(u + \ddot{q}^* - L\delta\dot{q} - P\delta q) + \hat{B}(q, \dot{q})\dot{q} + \hat{C}(q). \tag{6}$$

The role played by the external signal u should be that of acting as a correcting term allowing the perfect tracking of a given reference trajectory $q^*(t)$, $t \in [0, T]$, even in the presence of, possibly very high, mismatches between the various model terms and its estimates.

Given the reference motion $q^*(t)$, denote with $u^*(t)$ the corresponding unique (unknown) signal allowing its perfect tracking under the assumption of null initial-error conditions. Then, the problem will be just that of devising a suitable procedure to compute the unknown control function u^* without increasing the complexity of the underlying regulator structure (6).

In order to approach such a problem, it may be convenient to substitute the regulator espression (6) into the real system model (1), thus, obtaining the corresponding closed-loop equations

$$\begin{cases} \hat{A}^{-1}(q)[A(q)\ddot{q} + \bar{B}(q, \dot{q})\dot{q} + \bar{C}(q)] - \ddot{q}^* + L\delta\dot{q} + P\delta q = u \\ \delta q(0) = \delta\dot{q}(0) = 0 \end{cases} \tag{7}$$

where the nonsingularity of $\hat{A}(q)$ has been assumed and, for ease of notations, the following quantities have been defined

$$\begin{cases} \bar{B}(q, \dot{q}) \triangleq B(q, \dot{q}) - \hat{B}(q, \dot{q}) \\ \bar{C}(q) \triangleq C(q) - \hat{C}(q). \end{cases} \tag{8}$$

Then, according with Kawato et al. [10], by denoting with $q = T(u)$ the operator corresponding to the "solution process" of (7) (that is the one relating the space of the input functions u with that of the solution functions q of (7), for $\delta q(0) = \delta \dot{q}(0) = 0$), it can be trivially recognized that finding function u^* just corresponds to solve the functional equation

$$q^* = T(u) \tag{9}$$

having u^* as unique solution.

At this point the reader should be, however, advised that the simple references to the "operator theory" that will be made in the following, actually will not go far beyond this section; in fact, they will be instrumentally used only as a powerful and very general mean for better explaining (in the authors' opinion) the rationale underlying the structure of the learning control procedure considered in this chapter.

Furthermore, the reader should also note that, although the operator in (9) (or equivalently the differential equation (7)) is partially unknown, the corresponding mapping (from u to q) is, however, always available to measurements. (This as a consequence of the experimental possibility of applying signals u to the real-world controlled system.)

With this in mind, reconsider the functional equation (9), and observe that a possible procedure that could lead to its soluton u^* is the so-called "Iterative Newton method" in function spaces. It's rationale will now briefly be recalled for the reader's convenience, but without any claim of theoretical completeness (see, for instance, Kantorovich et al. [13], or any good book on functional analysis, for theoretical insights in the subject).

Consider a tentative input signal u_o applied to the real system (7), together with the corresponding (measured and recorded) output q_o, both defined in the time interval $[0, T]$.

Provided u_o and q_o are "not so far" from the desired ones u^* and q^* (on the basis of some functional norm, which is not yet the case to specify), we can approximately write

$$0 = T(u^*) - q^* \cong T(u_o) + T'(u_o)(u^* - u_o) - q^* = (q_o - q^*) + T'(u_o)(u^* - u_o) =$$
$$= \delta q_o + T'(u_o)(u^* - u_o) \tag{10}$$

where, analogously to functions of real variables, $T(u^*)$ has been approximated with its "functional Taylor expansion," computed at the "functional point" u_o, q_o, but arrested at the first term. In (10), the term $T'(u_o)$ represents the so-called (Frechet) "derivative" of the operator $T(u)$ evaluated at the functional point u_o, q_o. This term actually identifies a "linear operator" $T'(u_o)v$ having v as input function argument, specifically coinciding with $(u^* - u_o)$ in (10). As it can be easily proven via functional analysis considerations (but also argued on an intuitive basis), linear operator $T'(u_o)v$ simply corresponds to the "solution process" of the linear dynamic system (having $u^* - u_o$ as input an signal), just obtained by linearizing (7) around the motion defined by u_o, q_o.

More specifically, as it can be easily verified, such a linearized system takes on the form

$$\begin{cases} \hat{A}^{-1}(q_o)A(q_o)\ddot{\mu} + L\dot{\mu} + P\mu + [M(q_o, \dot{q}_o)\dot{\mu} + N(q_o, \dot{q}_o, \ddot{q}_o)\mu] = v \\ \mu(0) = \dot{\mu}(0) = 0 \end{cases} \quad (11)$$

where

$$M(q_o, \dot{q}_o) \triangleq \frac{\partial}{\partial \dot{q}} [\hat{A}^{-1}(q)\bar{B}(q, \dot{q})\dot{q}]_{q_o, \dot{q}_o} \quad (12a)$$

$$N(q_o, \dot{q}_o, \ddot{q}_o) \triangleq \frac{\partial}{\partial q} \{\hat{A}^{-1}(q)[A(q)\ddot{q} + \bar{B}(q, \dot{q})\dot{q} + \bar{C}(q)]\}_{q_o, \dot{q}_o, \ddot{q}_o} \quad (12b)$$

At this point, by resorting to the approximate functional equality (10), we may think to obtain an "estimate" u_1 of the unknown function u^* by simply functionally inverting (10), thus, obtaining

$$u_1 = u_o - \delta u_1 \quad (13)$$

where

$$\delta u_1 \triangleq T'^{-1}(u_o)\delta q_o \quad (14)$$

In the above espression (14), the term $T'^{-1}(u_o)$ identifies a linear operator which is simply the "inverse" of that identified by $T'(u_o)$. More specifically, it admits function δq_o as an input argument and performs a functional mapping, which is just opposite to that provided by $T'(u_o)$. Then, as a consequence of this fact, and also keeping into account that $T'(u_o)v$ was representing the "solution process" of (11), it is not difficult to realize that $T'(u_o)\delta q_o$ actually coincides with the inverse relationship.

$$\delta u_1 = \hat{A}^{-1}(q_o)A(q_o)\delta\ddot{q}_o + L\delta\dot{q}_o + P\delta q_o + [M(q_o, \dot{q}_o)\delta\dot{q}_o \\ + N(q_o, \dot{q}_o, \ddot{q}_o)\delta q_o], \quad (15)$$

which in turn, results an algebraic expression provided that, other than δq_o, $\delta\dot{q}_o$, and $\delta\ddot{q}_o$ were recorded in order to be given as input arguments to the right-hand side of (15).

Once the function estimate u_1 has been obtained from (13), via (15), it also can be applied to the true system (7) (with the same null initial-error conditions) in order to measure and record its corresponding new outputs q_1, $\dot{q}_1$, and $\ddot{q}_1$.

These, and u_1 itself, can, in turn, be used as a new functional starting point for repeating the same reasonings as before.

In this way, a recursive generalization of the procedure can be obtained, taking on the form

$$\begin{cases} u_{i+1} = u_i - \delta u_{i+1} \\ \delta u_{i+1} = T'^{-1}(u_i)\delta q_i \end{cases} \quad i = 0, 1, \ldots \tag{16}$$

where obviously $T'^{-1}(u_i)\delta q_i$ has the same algebraic form as the right-hand side (15), written with respect to trial index i.

The recursive procedure (16) represents the previously mentioned iterative Newton method in function spaces that, provided it converges, guarantees the solution of the original functional equation (9).

The fact that it appears to be structured as a "learning process" is noteworthy, in the sense that it explicitly uses the trajectory-error informations acquired during the execution of a motion trial, as corrective terms for the next trial, with the aim of improving it.

However, despite this fact, the learning procedure (16) does not yet completely match the philosophy of this chapter, because it still requires a large amount of informations about the controlled system (i.e., the knowledge of the terms relevant to its linearization).

To overcome this drawback, a possibility is offered by the reasonable idea of considering a simplified version of (16), namely the one given by

$$\begin{cases} u_{i+1} = u_i - \delta u_{i+1} \\ \delta u_{i+1} = \Gamma\delta q_i \end{cases} \quad i = 0, 1, \ldots \tag{17}$$

where the linear operator $\Gamma\delta q_i$ now corresponds to the time invariant algebraic relationship

$$\delta u_{i+1} = \delta \ddot{q}_i + L\delta q_i + P\delta q_i \tag{18}$$

which can be interpreted (by direct comparison) as an "approximation" of the more complex relationships (15), written for the generic trial index i.

The iterative relationships (17) and (18) actually represent the particular kind of learning procedures that will be considered throughout this chapter.

Before concluding this section, it may be interesting to note how the approximation introduced by the use of (18) heavily depends on the choices of L and P, other than on the accuracy of the estimated inertia matrix $\hat{A}(q)$.

The structures and the values assigned to L and P, and the characteristics of the estimate $\hat{A}(q)$ will actually play an important role in the stability and convergence analysis investigations performed in the next two sections.

3.   GENERAL BOUNDEDNESS RESULTS

This section has to be considered as preliminary, since it is devoted to the achievement of some important results concerning trajectory error boundedness. Such results will be, in turn, widely used in Section four, where the stability and convergence analysis of the learning procedure will be carried out.

To this end, recall (and renumber for ease of exposition) the closed-loop equation (7) of the manipulator, namely

$$\begin{cases} \hat{A}^{-1}(A\ddot{q} + \bar{B}\dot{q} + \bar{C}) + L\delta\dot{q} + P\delta q - \ddot{q}^* = u \\ \delta q(0) = \delta\dot{q}(0) = 0 \end{cases} \tag{19}$$

where, the various matrices' and vectors' arguments have been dropped for the sake of simplicity.

For the above system note that, due to the assumed null initial conditions, the forcing signal u^* allowing the manipulator to follow the desired trajectory turns out to be

$$u^* = [\hat{A}^*]^{-1}(A^*\ddot{q}^* + \bar{B}^*\dot{q}^* + \bar{C}^*) - \ddot{q}^* \tag{20}$$

where, for ease of notations

$$A^* \triangleq A(q^*), \qquad \bar{B}^* \triangleq \bar{B}(q^*, \dot{q}^*), \qquad \bar{C}^* \triangleq \bar{C}(q^*), \text{ and } \qquad [\hat{A}^*]^{-1} \triangleq \hat{A}^{-1}(q^*).$$

Then, with reference to a generic signal u, and defining the input signal error δu^* as

$$\delta u^* \triangleq u - u^* \tag{21}$$

it is immediate to get (by subtracting u^* given by (20) from both the sides of (19))

$$\hat{A}^{-1}(A\delta\ddot{q} + \bar{B}\delta\dot{q}) + \hat{A}^{-1}(A\ddot{q}^* + \bar{B}\dot{q}^* + \bar{C}) + \\ + L\delta\dot{q} + P\delta q - [\hat{A}^*]^{-1}(A^*\ddot{q}^* + \bar{B}^*\dot{q}^* + \bar{C}^*) = \delta u^* \tag{22}$$

which represents the dynamic equation relating the trajectory errors with the corresponding input errors.

The solution of (22) as an algebraic equation with respect to $\delta\ddot{q}$ yields

$$\begin{aligned} \delta\ddot{q} &= A^{-1}\hat{A}[\delta u^* - L\delta\dot{q} - P\delta q + [\hat{A}^*]^{-1}(A^*\ddot{q}^* + \bar{B}^*\dot{q}^* + \bar{C}^*) \\ &\quad + -\hat{A}^{-1}(A\ddot{q}^* + \bar{B}q^* + \bar{C}) + \hat{A}^{-1}\bar{B}\delta\dot{q}] \\ &= A^{-1}\hat{A}[\delta u^* - L\delta\dot{q} - P\delta q + f(\delta q, \delta\dot{q})] \end{aligned} \tag{23}$$

where

$$f(\delta q, \delta \dot{q}) \triangleq [\hat{A}^*]^{-1}(A^*\ddot{q}^* + \bar{B}^*\dot{q}^* + \bar{C}^*) - \hat{A}^{-1}(A\ddot{q}^* + \bar{B}\dot{q}^* + \bar{C}) - \hat{A}^{-1}\bar{B}\delta\dot{q} \qquad (24)$$

is a continuous function in the space $X \triangleq [\delta q, \delta \dot{q}]$, Lipschitzian inside any bounded domain.

Moreover, by reconsidering (22), a little algebra (namely adding $\delta\ddot{q}$ to both sides of (22) shows that it can be rewritten as

$$\delta\ddot{q} + L\delta\dot{q} + P\delta q = \delta u^* + (I - \hat{A}^{-1}A)\delta\ddot{q} + f(\delta q, \delta\dot{q}). \qquad (25)$$

Then, as a consequence, the substitution of (23) in the right-hand side of (25) leads to the dynamic relationship

$$\delta\ddot{q} + L\delta\dot{q} + P\delta q = A^{-1}\hat{A}\delta u^* - (A^{-1}\hat{A} - I)(L\delta\dot{q} + P\delta q) + A^{-1}\hat{A} f(\delta q, \delta\dot{q}) \qquad (26)$$

which can also be interpreted as a linear time-invariant system, having dynamic behavior characterized by L and P, driven by a combination of the forcing signal δu^* and complex state-dependent perturbation terms.

Equation (26) will actually play an essential role within the analysis performed in both this and the following section.

At this point, before entering into the details of successive developments, let us first assume (with a little loss of generality) that matrices L and P can be assigned having the structure

$$L = \left(2\alpha + \frac{1}{2} \right) I; \qquad P = \alpha I, \qquad (27)$$

being α is a non-negative scalar parameter.

Moreover, let us also recall the following property, whose proof can be found in the Appendix.

Property 1. for a linear MIMO time-invariant decoupled system having the form

$$\delta\ddot{q} + \left(2\alpha + \frac{1}{2} \right) \delta\dot{q} + \alpha\delta q = U \qquad \left(\alpha > \frac{1}{4} \right) \qquad (28)$$

and subject to null initial conditions, the following hold true

$$\begin{cases} \|\delta q\|_t \leq \gamma_p(t, \alpha)\|U\|_t \\ \|\delta\dot{q}\|_t \leq \gamma_v(t, \alpha)\|U\|_{it} \\ \left\| \left(2\alpha + \frac{1}{2} \right) \delta\dot{q} + \alpha\delta q \right\|_t \leq g(t, \alpha)\|U\|_t \end{cases} \qquad (29)$$

where scalars $\gamma_p(t, \alpha)$, $\gamma_v(t, \alpha)$, and $g(t, \alpha)$ are continuous non-negative, increasing time functions satisfying the conditions

$$\gamma_p(0, \alpha) = \gamma_v(0, \alpha) = g(0, \alpha) = 0 \qquad \forall \alpha$$

$$\lim_{t \to \infty} \{\gamma_p(t, \alpha)\} \leq \frac{1}{\alpha}$$

$$\lim_{t \to \infty} \{\gamma_v(t, \alpha)\} \leq \frac{1}{\alpha}$$

$$\lim_{t \to \infty} \{g(t, \alpha)\} \leq 1 + \frac{1}{2\alpha} \qquad\qquad (30)$$

$$\lim_{\alpha \to \infty} \gamma_p(t, \alpha) = \lim_{\alpha \to \infty} \gamma_v(t, \alpha) = 0 \qquad \forall\, t \geq 0$$

$$\lim_{\alpha \to \infty} g(t, \alpha) = 1 \qquad\qquad \forall\, t > 0$$

and where the following euclidean norm is used in correspondence of any time-varying vector $x(t) \in \mathcal{R}^n$

$$\|x\|_t \triangleq \sup_{0 \leq \tau \leq t} \|x(\tau)\| \qquad\qquad (31)$$

With this in mind, let us now reconsider Equation (26) in correspondence of any given $\alpha > 1/4$ in (27) and input signal error $\delta u^*(t)$. Then, observe that, in the worst case, due to the assumed null initial conditions, $\|\delta q\|_t$ and $\|\delta \dot{q}\|_t$ will be monotonically increasing continuous time functions, starting from zero, thus, implying

$$\forall \epsilon > 0 \quad \exists\, t^* > 0, \quad \text{such that} \begin{cases} \|\delta q\|_t \leq \epsilon \\ \|\delta \dot{q}\|_t \leq \epsilon \end{cases} \quad t \in [0,\, t^*] \qquad (32)$$

where t^* is the first time-instant where an equality sign holds (if any). As a consequence of the above fact, and by defining

$$U = A^{-1}\hat{A}\delta u^* - (A^{-1}\hat{A} - I)(L\delta\dot{q} + P\delta q) + A^{-1}\hat{A}f(\delta q, \delta\dot{q}) \qquad (33)$$

in the interval $[0,\, t^*]$ the following holds true

$$\|U\|_t \leq \|A^{-1}\hat{A}\|\|\delta u^*\|_t + \|A^{-1}\hat{A} - I\|\left(2\alpha + \frac{1}{2}\right)\delta\dot{q} + \alpha\delta q\|_t +$$

$$+ \beta(\epsilon)max\{\|\delta q\|_t, \|\delta\dot{q}\|_t\} \qquad t \in [0,\, t^*] \qquad (34)$$

where $\beta(\epsilon)$ represents the global Lipschitz constant of $A^{-1}\hat{A}f(\delta q, \delta\dot{q})$, in the domain characterized by $\|\delta q\|_t \leq \epsilon$, $\|\delta\dot{q}\|_t \leq \epsilon$.

Moreover, let us define

$$\gamma(t, \alpha) \triangleq max\{\gamma_p(t, \alpha), \gamma_v(t, \alpha)\} \tag{35}$$

$$\eta \triangleq \max_q \{\|A^{-1}(q)\hat{A}(q) - I\|\} \tag{36}$$

$$\lambda \triangleq \max_q \{\|A^{-1}(q)\hat{A}(q)\|\} \tag{37}$$

where in (36) and (37) the boundedness of matrix $\hat{A}(q)$ has been assumed. Then, upper bounding the second and third term of the right-hand side of (34) on the basis of property one and the above definitions, we can get the new condition

$$\|U\|_t \leq \lambda\|\delta u^*\|_t = [\eta g(t, \alpha) + \beta(\epsilon)\gamma(t, \alpha)]\|U\|_t \qquad t \in [0, t^*] \tag{38}$$

It is then apparent from the above condition that it can lead to an explicit upper bound for $\|U\|_t$ only in correspondence of all the times where the second term in brackets is less than one.

By keeping into account that $\gamma(t, \alpha)$ and $g(t, \alpha)$ are actually continuous functions starting from zero, it is an easy matter to see that a time interval $[0, \tilde{t}]$ certainly exists, such that

$$\eta g(t, \alpha) + \beta(\epsilon)\gamma(t, \alpha) \leq 1 \qquad t \in [0, \tilde{t}] \tag{39}$$

with the equality holding for $t = \tilde{t}$. This in turn assures that, within the time interval $[0, \bar{t})$, where

$$\bar{t} \triangleq min\{\tilde{t}, t^*\} \tag{40}$$

it will certainly hold

$$\|U\|_t \leq \frac{\lambda\|\delta u^*\|_t}{1 - \eta g(t, \alpha) - \beta(\epsilon)\gamma(t, \alpha)} \qquad t \in [0, \bar{t}) \tag{41}$$

At this point, by again applying Property one, it follows that

$$\left.\begin{matrix} \|\delta q\|_t \\ \|\delta\dot{q}\|_t \end{matrix}\right\} \leq \gamma(t, \alpha)\|U\|_t \leq \frac{\|\delta u^*\|_t \lambda\gamma(t, \alpha)}{1 - \eta g(t, \alpha) - \beta(\epsilon)\gamma(t, \alpha)} \qquad t \in [0, \bar{t}) \tag{42a}$$

$$\left\|\left(2\alpha + \frac{1}{2}\right)\delta\dot{q} + \alpha\delta q\right\|_t \leq \frac{\|\delta u^*\|_t \lambda g(t, \alpha)}{1 - \eta g(t, \alpha) - \beta(\epsilon)\gamma(t, \alpha)} \qquad t \in [0, \bar{t}) \tag{42b}$$

where the last terms in both (42a and b) are continuous increasing time functions starting from zero.

Up to this point, it has been, therefore, shown that, for any given $\alpha > \frac{1}{4}$ and $\delta u^*(t)$, there always exists an initial subinterval $[0, \tilde{t})$ inside $[0, t^*]$, where position and velocity errors are norm bounded by an assigned ϵ, such that the behavior of both $\|\delta q\|_t$, $\|\delta \dot{q}\|_t$ is, at worst, characterized by the increasing and starting from zero time function appearing in the right-hand side of (24b).

Nevertheless, no information about the size of the above introduced subinterval $[0, \tilde{t}]$ has yet been provided. In order to get such information, it is sufficient to consider the right-hand side of (24a) inserted within the following inequality condition

$$0 \le \frac{\|\delta u^*\|_t \lambda \gamma(t, \alpha)}{1 - \eta\, g(t, \alpha) - \beta(\epsilon)\gamma(t, \alpha)} \le \epsilon \qquad (43)$$

and, consequently, show that its solution with respect to t actually gives a lower bound for $\tilde{t} \le t^*$.

To do this, first note that the (starting from zero and increasing) time function considered in (43), exhibits a vertical behavior for t approaching $\tilde{t}$ (defined by (39)), thus, necessarily implying that it attains the ϵ value at a time instant $\hat{t} < \tilde{t}$.

Then, keeping into account definition (40), it has only to be shown that, also, inequality $\hat{t} \le t^*$ holds. To this aim, assume by contradiction $t^* < \hat{t}$. This would actually imply, due to (24b) and the increasing behavior of right-hand side of (43), that

$$\max\{\|\delta q\|_t^*, \|\delta \dot{q}\|_t^*\} < \epsilon$$

which would contradict the assumption that in t^* at least one of $\|\delta q\|_t$, $\|\delta \dot{q}\|_t$ attains the ϵ value.

On the basis of all the previous considerations, it is now possible to state and successively prove the following basic theorem.

Theorem 1. given a time interval $[0, T]$, not necessarily finite, consider the class S of input error signals δu^* having norm ranging within a common bound ξ in $[0, T]$. Then, provided

$$\eta \triangleq \max_q\{\|A^{-1}(q)\hat{A}(q) - I\|\} < 1 - \sigma \qquad \sigma \in (0, 1] \qquad (44)$$

the following properties hold true

$$\forall \epsilon > 0 \qquad \exists \bar{\alpha}_\epsilon > \frac{1}{4}, \text{ such that} \qquad \alpha > \bar{\alpha}_\epsilon \Rightarrow$$

$$\Rightarrow \begin{cases} \left. \begin{array}{l} \|\delta q\|_t \\ \|\delta \dot{q}\|_t \end{array} \right\} \le \dfrac{\|\delta u^*\|_t \lambda \gamma(t, \alpha)}{1 - \eta\, g(t, \alpha) - \beta(\epsilon)\lambda(t, \alpha)} \le \epsilon & \quad (45\text{a}) \\[4em] & \quad \begin{array}{l} \forall t \in [0,\ T] \\ \forall \delta u^* \in S \end{array} \\[2em] \left\|\left(2\alpha + \dfrac{1}{2}\right)\delta\dot{q} + \alpha\delta q\right\|_t \le \dfrac{\|\delta u^*\|_t \lambda\, g(t, \alpha)}{1 - \eta\, g(t, \alpha) - \beta(\epsilon)\gamma(t, \alpha)} & \quad (45\text{b}) \end{cases}$$

Proof—it is apparent from the previous discussion that conditions (45a and b) are proven whenever condition (43) is shown to hold for any $\delta u^* \in S$ within the given time interval $[0,\ T]$. As it can be easily verified (by using ξ in upper bounding the central part of (43)), this is actually the case provided a suitable choice of α exists, forcing the fulfillment of the following condition

$$0 \le \frac{\xi\lambda\gamma(T, \alpha)}{1 - \eta\, g(T, \alpha) - \beta(\epsilon)\lambda(T, \alpha)} \le \epsilon \qquad (46)$$

To this end, note that, by virtue of Property one

$$1 - \eta g(T, \alpha) - \beta(\epsilon)\gamma(T, \alpha) > 1 - \eta\left(1 + \frac{1}{2\alpha}\right) - \beta(\epsilon)\gamma(T, \alpha), \qquad (47)$$

thus, implying that if an α exists, such that

$$0 \le \frac{\xi\lambda\gamma(T, \alpha)}{1 - \eta\left(1 + \dfrac{1}{2\alpha}\right) - \beta(\epsilon)\gamma(T, \alpha)} \le \epsilon, \qquad (48)$$

then, for the same choice of α also (46) turns out to be satisfied. Actually, (48) can be rewritten as

$$\xi\lambda\gamma(T, \alpha) \le \epsilon\left[1 - \eta\left(1 + \frac{1}{2\alpha}\right) - \beta(\epsilon)\gamma(T, \alpha)\right] \qquad (49)$$

At this point, by noting that the left-hand side of (49) can be made arbitrarily small for increasing α (due to Property one), whereas

$$\lim_{\alpha\to\infty} \epsilon\left[1 - \eta\left(1 + \frac{1}{2\alpha}\right) - \beta(\epsilon)\gamma(T, \alpha)\right] = \epsilon(1 - \eta) \qquad (50)$$

it is apparent that if η satisfies the condition of the theorem, then a value of α exists, such that (49) is satisfied, in turn, implying the fulfillment of (46).

Finally, note that the above reasoning line also applies if $T \rightarrow \infty$, thus, completely proving the theorem.   □

Remark 3.1. the above theorem actually provides a mild condition on the precision of the estimate $\hat{A}(q)$ of the inertia matrix which is sufficient for guaranteeing the possibility of achieving, for increasing α, any desired tracking accuracy, within any time interval and in correspondence of any norm-bounded class of input error signals.

Even if the theorem has been introduced in order to be widely referred in the next section, within the context of stability and convergence analysis of the learning procedure, it exhibits its own importance, since it clarifies the role that can be played by linear-state feedbacks within the "approximated computed torque" control schemes.   □

Remark 3.2. as a consequence of the results expressed in Theorem one, it appears that the key role is played only by the estimated inertia matrix $\hat{A}(q)$, since the effects of $\tilde{B}(q, \dot{q})$ and $\tilde{C}(q)$ (condensed in the Lipschitz constant $\beta(\epsilon)$) tend to vanish for increasing α.

Roughly speaking, it is, however, clear that more accurate estimates $\hat{B}$ and $\hat{C}$ tend to decrease the Lipschitz constant $\beta(\epsilon)$, thus, allowing smaller values of α guaranteeing the same tracking accuracy.

4.   BOUNDEDNESS AND CONVERGENCE ANALYSIS

The problem of proving the effectiveness of the learning procedure outlined in Section two and described by Equations (16), (17) will be dealt with in this section. More specifically, a wide use of the stability results of Section 3 will be made, in order to consider at the same time both the following fundamentals items: (1) uniform boundedness of the errors at any trial, and (2) existence of a contraction mapping assuring their convergence to zero for increasing number of trials.

To this aim, start by recalling that the updating law (17) for u_i, with the assumed choice of L and P, turns out to be

$$u_{i+1} = u_i - \delta\ddot{q}_i - \left(2\alpha + \frac{1}{2} \right) \delta\dot{q}_i - \alpha\delta q_i \qquad (51)$$

By subtracting the unknown signal u^* from both the sides of (51) and taking into account the definition of δu^*, given in Section three, it is immediate to get

$$\delta^*_{i+1} = \delta u^*_i - \delta\ddot{q}_i - \left(2\alpha + \frac{1}{2} \right) \delta\dot{q}_i - \alpha\delta q_i, \qquad (52)$$

thus, trivially showing that the same recursive relationship as (51) also holds for the unknown "input signal error."

For the purpose of this section, first rewrite the closed-loop equation (26) of the manipulator relevant to the generic i-th trial of movement, in the form

$$\delta\ddot{q}_i + \left(2\alpha + \frac{1}{2}\right)\delta\dot{q}_i + \alpha\delta q_i = \delta u_i^* + (A_i^{-1}\hat{A}_i - I)(\delta u_i^* - \left(2\alpha + \frac{1}{2}\right)\delta\dot{q}_i - \alpha\delta q_i)$$
$$+ A_i^{-1}\hat{A}_i f(\delta q_i, \delta\dot{q}_i) \tag{53}$$

where δu_i^* has been added and subtracted, and where, obviously,

$$A_i^{-1} \triangleq A^{-1}(q_i), \text{ and } \hat{A}_i \triangleq \hat{A}(q_i).$$

Then, a direct comparison of the right-hand side of (52) with the left-hand side of (53) yields

$$\delta u_{i+1}^* = -(A_i^{-1}\hat{A}_i - I)(\delta u_i^* - \left(2\alpha + \frac{1}{2}\right)\delta\dot{q}_i - \alpha\delta q_i) - A_i^{-1}\hat{A}_i f(\delta q_i, \delta\dot{q}_i) \tag{54}$$

which expresses the recursive dependence of the input error signal at the $(i + 1)$-th trial on the input error signal and on the nonlinear terms characterizing the system dynamic at the i-th trial.

Remark 4.1. the nature of the recursive dependence (54), generalized to the case of unstructured feedback matrices L and P (i.e.,

$$\delta u_{i+1}^* = -(A_i^{-1}\hat{A}_i - I)(\delta u_i^* - L\delta\dot{q}_i - P\delta q_i) - A_i^{-1}\hat{A}_i f(\delta q_i, \delta\dot{q}_i)$$

may allow a better understanding of the motivations underlying the structure and the kind of analysis performed in this chapter.

First, it can be remarked that the above recursive relation, being a direct consequence of the general structure of the learning process defined in the second section, is actually a very general one. In fact, similar relationships appear within almost all the works concerning with learning procedures applied to non-linear (commonly robotic) systems.

Many approaches to the analysis of such procedures start from relationships similar to the above one and (after definition of suitable operator norm for trajectory and input signal errors) develop on the basis of the implications contained in the well-known 'Gronwall's Lemma' (applied to the above relationship, in connection to the dynamic equation (22) or the equivalent (25), (26)).

Actually, a straightforward application of the Gronwall's Lemma requires the fulfillment of global Lipschitz conditions by all the terms appearing in both the error-system equation and the above recursive dependence. It is, however, a matter of fact that such global Lipschitz conditions cannot generally be invoked when referring to robotized mechanical systems. In fact, this is due to the presence of the term $f(\delta q, \delta \dot{q})$, which, in turn, embeds the term $\bar{B}(q, \dot{q})\delta \dot{q}$, known to be a quadratic (thus, not globally Lipschitzian) vector function of the velocity error $\delta \dot{q}$ (the reader can easily verify this fact by referring to the structure of matrix $B(q, \dot{q})$ recalled in expression (81) of Section 5).

In order to overcome the above problem, it is then clear that the previous achievement of conditions guaranteeing the uniform boundedness of position and velocity errors, within all the learning trials, is obviously required. This necessity actually represents the reason of the effort spent in Section three, in order to get some preliminary general boundedness results.

Still, as a matter of fact, it also appears that once the results in Section three are available, a single proof guaranteeing both uniform boundedness at all the trials and the convergence to zero of the error signals can actually be provided, without invoking the use of the Gronwall's Lemma. (This will be described in the following of the present section.) □

At this point, the key result of this section can be immediately stated by means of the following.

Theorem 2. for any bounded reference trajectory q^*, $\dot{q}^*$, $\ddot{q}^*$, having duration T, not necessarily finite, provided

$$\eta \triangleq \max_q \{\|A^{-1}(q)\hat{A}(q) - I\|\} < \frac{1}{3} - \sigma \qquad \sigma \in \left(0, \frac{1}{3}\right) \tag{55}$$

then

$$\forall \epsilon > 0 \ \exists \ \bar{\alpha}_\epsilon > \frac{1}{4}, \text{ such that } \alpha \geq \bar{\alpha}_\epsilon \Rightarrow \begin{cases} \|\delta q_i\|_t \leq \epsilon \\ \|\delta \dot{q}_i\|_t \leq \epsilon \end{cases} \forall i; \qquad \forall t \in [0, T] \tag{56}$$

Moreover, a contraction mapping exists of the form

$$\begin{cases} \|\delta u_i^*\|_t \leq \mu^i(\alpha, t)\|\delta u_0^*\|_t & (57a) \\ \qquad\qquad\qquad\qquad \mu(\alpha, t) < 1 \qquad t \in [0, T] \\ \|\delta q_i\|_t \\ \|\delta \dot{q}_i\|_t \end{cases} \leq \mu^i(\alpha, t)\epsilon \qquad (57b)$$

where the contraction factor $\mu(\alpha, t)$ is defined by (62) below. This, in turn, guarantees the convergence of the learning procedure, i.e.,

$$\lim_{i \to \infty} \|\delta u_i^*\|_t = \lim_{i \to \infty} \|\delta q_i\|_t = \lim_{i \to \infty} \|\delta \dot{q}_i\|_t = 0 \qquad \forall t \in [0, T]$$

Proof. as already mentioned, it will be proven, by induction, the uniform boundedness of the input and trajectory error signals together with the existence of the mentioned contraction mapping.

To this end, start by considering the first trial of movement, carried out on the basis of a tentative signal u_o, norm bounded in $[0, T]$. Then, from the definition of δu^* we have

$$\|\delta u_o^*\|_t = \|u_o - u^*\|_t = \|u_o - [\hat{A}^*]^{-1}(A^*\ddot{q}^* + \bar{B}^*\dot{q}^* + \bar{C}^*) - \ddot{q}^*\|_t, \quad t \in [0, T] \qquad (58)$$

Moreover, let us fix a value, say ϵ, relevant to the bound we want to assign to $\|\delta q_o\|_t$ and $\|\delta \dot{q}_o\|_t$ in $[0, T]$.

Then, it is plain that, if $\|A^{-1}A - I\|$ satisfies condition (55) of the theorem (actually more restrictive than the one reported in Theorem 1), it is, however, assured on the basis of Theorem 1, that

$$\exists\ \delta_\epsilon > \frac{1}{4}\ \text{such that}\ \alpha > \bar{\alpha}_\epsilon\ \text{allows}$$

$$\begin{cases} \left.\begin{array}{r} \|\delta q_o\|_t \\ \|\delta \dot{q}_o\|_t \end{array}\right\} \leq \dfrac{\|\delta u_o^*\|_t\, \lambda\, \gamma(t, \alpha)}{1 - \eta\, g(t, \alpha) - \beta(\epsilon)\gamma(t, \alpha)} \leq \epsilon \\[1em] \left\|\left(2\alpha + \dfrac{1}{2}\right)\delta\dot{q}_o + \alpha\delta q_o\right\|_t \leq \dfrac{\|\delta u_o^*\|_t\lambda\, g(t, \alpha)}{1 - \eta\, g(t, \alpha) - \beta(\epsilon)\gamma(t, \alpha)} \end{cases} \qquad t \in [0, T] \qquad (59)$$

Refer now to the second trial and to the corresponding input error signal δu_1^*. Upper bounding its expression given by (54) for $i = 0$, we have

$$\|\delta u_1^*\|_t \leq \eta\|\delta u_o^*\|_t + \eta\left\|\left(2\alpha + \frac{1}{2}\right)\delta q_o + \alpha\delta\dot{q}_o\right\|_t$$

$$+ \beta(\epsilon)\max\{\|\delta q_o\|_t, \|\delta\dot{q}_o\|_t\} \qquad t \in [0, T] \qquad (60)$$

Then, taking into account inequalities (59), a little algebra leads to

$$\|\delta u_1^*\|_t \leq \|\delta u_o^*\|_t\left[\eta + \frac{[\eta\, g(t, \alpha) + \beta(\epsilon)\gamma(t, \alpha)]\lambda}{1 - [\eta\, g(t, \alpha) + \beta(\epsilon)\gamma(t, \alpha)]}\right] \qquad t \in [0, T] \qquad (61)$$

We shall now shortly investigate the possibility of imposing, at the first trial, values for α not only greater than $\bar{\alpha}_\epsilon$ assuring the validity of (59), but also such that

$$\mu(\alpha, t) \triangleq \eta + \frac{[\eta\, g(t, \alpha) + \beta(\epsilon)\gamma(t, \alpha)]\lambda}{1 - [\eta\, g(t, \alpha) + \beta(\epsilon)\gamma(t, \alpha)]} < 1, \qquad (62)$$

which would imply a contraction of the input error signal from the first to the second trial.

In order to prove the existence of such values, first note that the term in the left-hand side of (62) can be upper bounded by upper bounding λ as follows

$$\lambda \triangleq \max_q \{\|A^{-1}(q)\hat{A}(q)\|\} = \max_q \{\|A^{-1}(q)\hat{A}(q) - I + I\|\}$$

$$\leq 1 + \max_q \{\|A^{-1}(q)\hat{A}(q) - I\|\} = 1 + \eta,$$

thus, assuring that the fulfillment of the condition

$$\eta + \frac{[\eta \, g(t, \alpha) + \beta(\epsilon)\gamma(t, \alpha)][1 + \eta]}{1 - [\eta \, g(t, \alpha) + \beta(\epsilon)\gamma(t, \alpha)]} < 1 \tag{63}$$

will automatically imply the fulfillment of (62).

Then, in order to satisfy this last condition, consider the worst case for (63) to hold, namely (since the right-hand side of (63) is an increasing time function) when t equals T. By solving the inequality (63) with respect to η, a straightforward algebra leads to the condition

$$\eta < \frac{1 - 2\beta(\epsilon)\gamma(T, \alpha)}{1 + 2g(T, \alpha)} \tag{64}$$

Here again, since

$$\lim_{\alpha \to \infty} \frac{1 - 2\beta(\epsilon)\gamma(T, \alpha)}{1 + 2g(T, \alpha)} = \frac{1}{3} \tag{65}$$

it is apparent that if η lies within the bound specified in the statement of the theorem, then at the first trial it is always possible to choose α in order to assure both (59) and

$$\|\delta u_1^*\|_t \leq \mu \|\delta u_o^*\|_t \qquad t \in [0, T] \tag{66}$$

where, for ease of notation, the dependence on α and t in $\mu(\alpha, t)$ has been dropped.

Moreover, with such a choice, by virtue of Theorem one, it is also assured that

$$\left.\begin{array}{c} \|\delta q_1\|_t \\ \|\delta \dot{q}_1\|_t \end{array}\right\} \leq \mu \epsilon \qquad t \in [0, T] \tag{67}$$

Thus, the first step of the induction procedure has been proven. Now assume that (57a) and (57b) together with the structure reported in (59) hold at the i-th trial, for the same choice of α. Here again

$$\|\delta u_{i+1}^*\|_t \leq \eta \|\delta u_i^*\|_t + \eta \left\|\left(2\alpha + \frac{1}{2}\right)\delta \dot{q}_i + \alpha \delta q_i\right\|_t$$

$$+ \beta(\mu^i \epsilon)\max\{\|\delta q_i\|_t, \|\delta \dot{q}_i\|_t\} \qquad t \in [0, T] \tag{68}$$

At this point, if (57a) holds at the i-th trial, and since $\beta(\mu^i\epsilon) \leq \beta(\epsilon)$ (β represents a Lipschitz constant and μ is less than 1), equation (68) can be rewritten as

$$\|\delta u^*_{i+1}\|_t \leq \eta\mu^i\|\delta u^*_0\|_t + \eta\left\|\left(2\alpha + \frac{1}{2}\right)\delta\dot{q}_i + \alpha\delta q_i\right\|_t$$

$$+ \beta(\epsilon)\max\{\|\delta q_i\|_t, \|\delta\dot{q}_i\|_t\} \qquad t \in [0, T] \qquad (69)$$

Then, following the same reasoning line as before, it can be rewritten as

$$\|\delta u^*_{i+1}\|_t \leq \mu^i\|\delta u^*_0\|_t\left[\eta + \frac{[\eta\, g(t, \alpha) + \beta(\epsilon)\gamma(t, \alpha)](1 + \eta)}{1 - [\eta\, g(t, \alpha) + \beta(\epsilon)\gamma(t, \alpha)]}\right] \qquad t \in [0, T] \qquad (70)$$

Moreover, keeping into account the definition (62) for the contraction factor μ, which results to be independent of i, it is immediate to get

$$\|\delta u^*_{i+1}\|_t \leq \mu^{i+1}\|\delta u^*_o\|_t \qquad t \in [0, T] \qquad (71)$$

which in turn implies

$$\left.\begin{array}{c}\|\delta q_{i+1}\|_t\\\|\delta\dot{q}_{i+1}\|_t\end{array}\right\} \leq \mu^{i+1}\epsilon \qquad t \in [0, T] \qquad (72)$$

showing the existence of a contraction mapping of the form (57a), (57b).

Finally, note that the above reasoning line also apply if $T \to \infty$, thus, completely proving the theorem. $\qquad\square$

Remark 4.2. if compared with Theorem 1, Theorem 2 provides a little more restrictive, but still mild, condition on the accuracy of the estimate $\hat{A}(q)$, guaranteeing the possibility of achieving, not only uniform trajectory errors boundedness at any trial (within any specified bound as in Theorem 1), but also the uniform convergence to zero of the errors, with increasing number of trials.

The proof of the theorem clearly shows the need of restricting the accuracy condition on $\hat{A}(q)$, in order to guarantee the fulfilment of both the above items (uniform boundedness and convergence).

Remark 4.3. note that whenever convergence is guaranteed (by condition (55) and an appropriate choice of α, as indicated in Theorem 2) then, due to the existing contraction mapping, the convergence rate is established by the contraction factor $\mu(\alpha, t)$.

Since it is an increasing time function, it follows that convergence rates are decreasingly distributed along the time axis. More specifically, the greater is the considered time istant $t \in [0, T]$, the slower is the convergence to zero of the errors at that time.

Moreover, observe from (62) (upper bounded by (63)) that, whatever is $t \in$

$[0, T]$, the corresponding contraction factor is decreasing with η; thus, meaning that the better is the accuracy of $\hat{A}(q)$, the faster is the convergence of the procedure at any time.

Meanwhile, the contraction factor is also decreasing with the Lipschitz constant $\beta(\epsilon)$; thus, showing that also better accuracies of the estimates $\hat{B}(q, \dot{q})$ and $\hat{C}(q)$ contribute in accelerating convergence.

However, their contribution is in a sense of limited nature, since they cannot, in any case, force the contraction factor (again see (63)) to be lower than η (representing a lower bounding value), which can be decreased only by improving the accuracy of the estimated inertia matrix $\hat{A}(q)$.

Note, also, that similar considerations also apply when referring to the parameter α. More precisely, as it can be again seen from (62) upper bounded by (63), its contribution to an improvement of the contraction factor is actually limited, since, it also, cannot in any case force $\mu(\alpha, t)$ to be lower than η.

Nevertheless, the contribution of α is, however, important from a practical point of view. In fact (apart guaranteeing boundedness and convergence once it has been chosen beyond a certain threshold), the increasing of its value also allow to reduce the global error level ϵ at any trial; thus, contributing to a qualitative global improvement of the learning process (even if accelerating in a limited manner).

5.   IDENTIFICATION OF THE INERTIA PARAMETERS

The development of the previous sections have substantially shown the possibility of properly structuring a convergent Learning procedure, provided the inertia matrix is known with a degree of accuracy ranging within a specific bound.

The basic aim of the present section is, therefore, that of indicating a simple algorithm especially suited for the identification of the inertia parameters only, and useful for the attainment of the required accuracy condition if it cannot be a-priori guaranteed. To this end, start by considering the expression of the total work input to the mechanical system, at a given time instant, in absence of frictional forces, that is

$$\int_0^t \dot{q}'M d\tau - \int_0^t \dot{q}'C(q)d\tau = \frac{1}{2}\dot{q}'A(q)\dot{q} \qquad (73)$$

("'" indicates the transpose), where the two terms of the left-hand side represent the work produced by the torque vector M and the gravitational forces $C(q)$, respectively (each one expressed as the time integral of the corresponding power), while the term in the right-hand side represent the total kinetic energy (which coincides with its variation under the assumption of null velocity at the initial time).

From this point on, matrix $A(q)$ and vector $C(q)$ will actually be assumed known in their internal "structure," and, moreover, the possibility of exploiting such knowledge, for identification purposes only, will be supposed. Then by keeping into account that matrix $A(q)$ and vector $C(q)$ have elements which are linear in their (unknown but constant) parameters, it immediately follows that (73) can also be rewritten as

$$\int_0^t \dot{q}'M d\tau - \left[\int_0^t \bar{h}_2'(q, \dot{q})d\tau\right]\theta_2 = h_1'(q, \dot{q})\theta_1 \qquad (74)$$

where θ_1 and θ_2 are the minimal dimension (unknown) parameter vectors characterizing $A(q)$ and $C(q)$, respectively, while $h_1'(q, \dot{q})$, and $\bar{h}_2'(q, \dot{q})$ are row vectors whose elements, as a consequence of the previous assumptions, are known functions of their arguments. Note also that $h_1'(q, \dot{q})$, $\bar{h}_2'(q, \dot{q})$, and the first term of the left hand side of (74) are on-line measurable. To simplify notations rewrite (74) as

$$y(t) = h'(t)\theta \qquad (75)$$

where

$$y(t) \triangleq \int_0^t \dot{q}'M d\tau; \qquad \theta \triangleq [\theta_1' : \theta_2']; \qquad h'(t) \triangleq [h_1'(t) : h_2'(t)];$$

and

$$h_1'(t) \triangleq h_1'(q(t), \dot{q}(t)); \qquad h_2'(t) \triangleq \int_0^t \bar{h}_2'(q, \dot{q})d\tau$$

From the structure of (75) it can be immediately seen that the discrete-time recursive least-squares algorithm could be used to identify θ with guaranteed convergence, provided $h'(t)$ exhibits a "persistently exciting" behavior.

Nevertheless, in this section we shall also suggest a continuous time version of recursive least-squares estimation for θ, which can be easily obtained as described in the following.

By premultiplying both sides of (75) by $h(t)$, integrating and dividing by t, we have

$$\frac{1}{t}\int_0^t h(\tau)y(\tau)d\tau = \frac{1}{t}\left[\int_0^t h(\tau)h'(\tau)d\tau\right]\theta \qquad (76)$$

that can be simply rewritten as, (with obvious meaning of notations)

$$z(t) = R(t)\theta \tag{77}$$

with $R(t) > 0$ almost surely, provided a sufficient excitation is input to the system. Note, since $h(t)h'(t) \geq 0$, we have divided by t in Equation (76) as a possible mean to assure the boundedness of $R(t)$.

Under this condition, a suitable asymptotic estimation for θ results to be the following

$$\dot{\hat{\theta}} = -R(t)\hat{\theta} + z(t) \tag{78}$$

the corresponding error equation ($e = \theta - \hat{\theta}$) being

$$\dot{e} = -R(t)e \tag{79}$$

For such a differential relationship, the candidate Lyapunov function

$$V(e) = \frac{1}{2} e'e \tag{80}$$

yields to

$$\dot{V}(e) = -e'R(t)e < 0 \tag{80}$$

which implies the convergence of $\hat{\theta}$ to the true parameters value and, as a consequence, the eventual fulfillment of the accuracy condition for the estimated inertia matrix $\hat{A}(q)$ before the achievement of the exact identification.

It must be explicitly remarked that the above algorithm, even if it produces an estimate also for vector $C(q)$ (which is not strictly required for learning purposes), is, however, characterized by a reduced dimensionality if compared with those used within the adaptive control field [11], [12], since matrix $B(q, \dot{q})$ (it is not required for learning purposes) is actually not identified. Moreover, still compared with those currently used within the adaptive control field [11], [12], the proposed algorithm also results to be simplified, since it is based on position and velocity measurements only.

Finally, it is also worth noting that matrix $B(q, \dot{q})$, even if not directly identified by the algorithm, can actually be deduced in an indirect way; this as a consequence of its well-known relation with matrix $A(q)$, given by

$$B(q,\dot{q}) = \begin{bmatrix} \dot{q}' \left\{ \frac{1}{2} \left[\frac{\partial a_1}{\partial q} + \left(\frac{\partial a_1}{\partial q} \right)' \right] + \frac{\partial}{\partial q^1} A(q) \right\} \\ \dot{q}' \left\{ \frac{1}{2} \left[\frac{\partial a_n}{\partial q} + \left(\frac{\partial a_n}{\partial q} \right)' \right] + \frac{\partial}{\partial q^n} A(q) \right\} \end{bmatrix} \tag{81}$$

with a_k, $k = 1, 2, \ldots, n$ denoting the k-th column of matrix $A(q)$ and q^k the k-th component of q.

More specifically, also assuming the structural knowledge of matrix $B(q, \dot{q})$, its convergence estimate is obtainable via a direct substitution of the components of $\hat{\theta}$ in appropriate positions within the matrix structure (81).

The simplest way of employement of the identification algorithm (with or without indirect identification of the $B(q, \dot{q})$ terms) within a learning context, is represented by an a-priori execution of some persistently exciting motions (even uncorrelated with the required one) allowing at least a partial identification of $A(q)$; then followed by the sequence of learning trials, converging toward the exact execution of the required motion.

A little more sophisticated, but more efficient use of the identification algorithm is instead represented by its embedding within the execution of the learning trials.

The presentation of a complete theory for this second case (currently under development) is, however, beyond the scope of this chapter, but it is mentioned here as a clear example of the very close emerging connection between learning and adaptive control research fields.

6. SIMULATION EXAMPLE

The growing literature in the field of learning processes, applied to robotic manipulators, has always provided many simulative examples showing the effectiveness of the procedures, at least from a practical point of view.

Within this chapter, which is, however, mainly of a theoretical nature, a simulative experience is also discussed for the sake of completeness. Nevertheless, we limit ourselves to the description of a single example, simple enough, but indicative of the potentiality of learning techniques. The description is also maintained at an indicative level. The interested reader is deferred to many existing papers for more detailed descriptions of obtained simulative results.

Consider, for instance, the simple case of a two-link planar manipulator having masses, inertia, and dimensions comparable to those of the human arm (Figure 2.1). Assume that the tip of the second link is asked to follow, in the plane, a trajectory similar to the figure "**a**," as it is also depicted in Figures 2.2, with prespecified velocity profiles within a time interval of three seconds.

Then, after off-line kinematic inversion, here performed on the basis of relationships,

$$\begin{cases} \dot{q} = J^{-1}(q)\dot{x} \\ q = \int_0^t \dot{q} \, d\tau \end{cases}$$

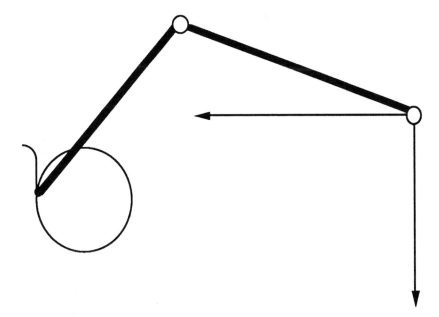

FIGURE 2.1.   Planar Robot Scheme with Reference Trajectory

with the Jacobian matrix (a purely geometric entity) assumed to be known, the joint-space reference trajectory $q^*(t)$, $\dot{q}^*(t)$ is obtained, and $\ddot{q}^*(t)$, consequently deduced on the basis of accurate off-line differentiation.

At this point, the simulation of the learning process has been performed (with $u_o = 0$), following the same theoretical guidelines as indicated in Section 4.

In particular, matrix $\hat{A}(q)$ has been chosen with the same structure of $A(q)$ but with modified parameters, however fulfilling condition (55) of Theorem 2. Estimated matrix $\hat{B}(q, \dot{q})$ has been chosen as the null one, while $\hat{C}(q)$ takes on a constant value equal to $C[q^*(0)]$. Moreover, feedback matrices L and P have been assigned the particular structure considered in this chapter, but the value of the parameter α has not been fixed very large, in order to have sufficiently high trajectory errors at the first trial.

As it can be seen, the simulative learning process exhibits a practical convergence within the first five to six trials. As it can also be argued from the reported figures, the greater the considered time instant within the considered time interval, the slower the convergence of the corresponding trajectory point.

Furthermore, as it was expected (even if not shown here) bettering the estimate inertia matrix $\hat{A}(q)$ greatly contributes in accelerating the convergence process, as indicated in Remark 4.3.

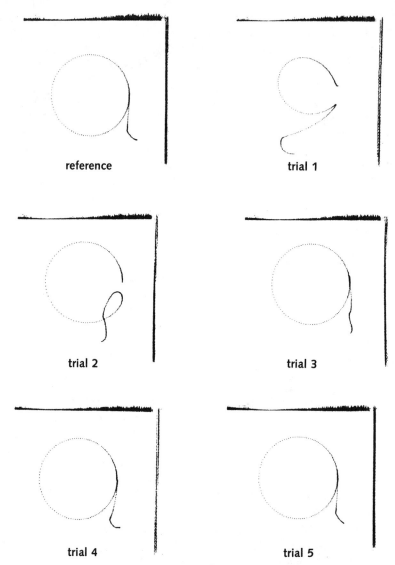

reference   trial 1

trial 2   trial 3

trial 4   trial 5

FIGURE 2.2.   Reference Trajectory and Set of Motion Trials

7. CONCLUSIONS AND FURTHER RESEARCH DIRECTIONS

Within this chapter, the general framework underlying the so-called Iterative Learning Control Techniques, applied to robotic systems, has been presented.

Sufficient conditions guaranteeing error boundedness and convergence have been provided and mainly expressed in terms of the accuracy required to the estimated inertia matrix of the system.

A simple algorithm for the identification of the inertia parameters, and useful for achieving the required accuracy conditions, has also been suggested. Such an algorithm establishes a possible close link between learning and adaptive techniques for robotic systems.

Finally, an indicative simulation example has been provided.

As to possible further research lines within the subject, we mention here only those that, in the authors' opinion, are considered the most important ones. First, recall the already mentioned need of exploring more into details the connections between learning and adaptive methodologies.

As a second point, it seems to be crucial to have a more deep investigation about the possibility of defining learning techniques for the control of manipulators directly performed in the cartesian or task space, thus, avoiding the separation between kinematic inversion and joint-space (learning) control (some preliminary considerations on this point can be found in [10]).

Third, it is worth mentioning the problem of devising learning possibilities within the so-called "point-to-point control problem": where only initial and final postures are assigned, and a joining trajectory must be specified minimizing some kind of cost functional (minimum total work, time, or others).

Last, but not least, the problem of learning within hybrid control conditions (integration of trajectory and contact force control during interaction with the environment) should also be considered with interest.

APPENDIX

Consider the SISO dynamic system described by the following differential equation

$$\ddot{y} + L\dot{y} + Py = u \qquad (A1)$$

subject to the null initial condition, and to the forcing signal u.

Assume that $L > 0$ and $P > 0$ can be assigned as arbitrary functions of a common parameter $\alpha > 0$; i.e.,

$$L = L(\alpha); \qquad P = P(\alpha)$$

The aim of this appendix is that of providing as suitable (minimal in some sense) structure for $L(\alpha)$ and $P(\alpha)$, in order to assign a desired behavior of the dynamic system (A1). In other words, for any forcing signal u, the system is asked to exhibit

$$\begin{cases} \|y\|_t \leq \gamma_p(t, \alpha)\|u\|_t \\ \|\dot{y}\|_t \leq \gamma_v(t, \alpha)\|u\|_t \end{cases} \qquad (A2)$$

with the operator norm defined for a scalar $x \in \mathfrak{R}$ as

$$\|x\|_t \triangleq \max_{0 \leq \tau \leq t} |x(\tau)| \tag{A3}$$

and with $\gamma_p(t, \alpha)$, $\gamma_v(t, \alpha)$ non-negative, increasing functions satisfying the conditions

$$\gamma_p(0, \alpha) = \gamma_v(0, \alpha) = 0 \qquad\qquad \forall \alpha \tag{A4a}$$

$$\lim_{t \to \infty} \{\gamma_p(t, \alpha)\} \leq \frac{1}{\alpha} \tag{A4b}$$

$$\lim_{t \to \infty} \{\gamma v(t, \alpha)\} \leq \frac{1}{\alpha} \tag{A4c}$$

$$\lim_{\alpha \to \infty} \gamma_p(t, \alpha) = \lim_{\alpha \to \infty} \gamma_v(t, \alpha) = 0 \qquad \forall t \geq 0 \tag{A4d}$$

To the above aim, first observe that $y(t)$ can be expressed as

$$y(t) = \int_0^t h(t - \tau)u(\tau)d\tau \tag{A5}$$

being $h(\cdot)$ the impulse response of (A1). Then, as a consequence

$$|y(t)| \leq \int_0^t |h(t - \tau)||u(\tau)|d\tau \tag{A6}$$

which, in turn, implies, since

$$\int_0^t |h(t - \tau)|d\tau = \int_0^t |h(\tau)|d\tau$$

$$\|y\|_t \leq \left(\int_0^t |h(\tau)|d\tau \right) \|u\|_t \tag{A7}$$

that can be rewritten as

$$\|y\|_t \leq \gamma_p(t, \alpha)\|u\|_t \tag{A8}$$

where, obviously

$$\gamma_p(t, \alpha) \triangleq \int_0^t |h(\tau)|d\tau \tag{A9}$$

is a non-negative, increasing function of time and such that $\gamma_p(0, \alpha) = 0$.

Moreover, as regard (A4b), it is an easy matter to see that, provided $L(\alpha)$ and $P(\alpha)$ are constrained to be such that

$$L^2(\alpha) - 4P(\alpha) > 0 \tag{A10}$$

(which implies (A1) to be an "overdamped" system), the following holds true

$$0 \leq h(\tau) = \frac{1}{r_1 - r_2} (e^{-r_2\tau} - e^{-r_1\tau}) \tag{A11}$$

with

$$r_1 \triangleq \frac{L(\alpha) + \sqrt{L^2(\alpha) - 4P(\alpha)}}{2}; \qquad r_2 \triangleq \frac{L(\alpha) - \sqrt{L^2(\alpha) - 4P(\alpha)}}{2} \tag{A12}$$

Consequently, from (A9)

$$\gamma_p(t, \alpha) \leq \lim_{t \to \infty} \gamma_p(t, \alpha) = \int_0^{+\infty} |h(\tau)| d\tau = \frac{US4}{1} \qquad \frac{1}{P(\alpha)} \tag{A13}$$

From (A13) it follows that the upper bound (A4b) can be satisfied by simply choosing

$$P(\alpha) = \alpha \tag{A14}$$

As to the analogous properties to be forced to hold also for $\|\dot{y}\|_t$, note that, similarly to (A8) and (A9), it turns out to be

$$\|y\|_t \leq \gamma_v(t, \alpha)\|u\|_t \tag{A15}$$

with

$$\gamma_v(t, \alpha) \triangleq \int_0^t |\dot{h}(\tau)| d\tau \tag{A16}$$

it also non-negative, increasing with time, and such that $\gamma_p(0, \alpha) = 0$. Moreover, due to the fact that

$$\int_0^{+\infty} \dot{h}(\tau) = \lim_{t \to \infty} h(t) = 0 \tag{A17}$$

and keeping into account that $\dot{h}(t)$ performs only one zero crossing, the use of (A16) and (A17) yields

$$\gamma_v(t, \alpha) \leq \lim_{t \to \infty} \gamma_v(t, \alpha) = \int_0^{+\infty} |\dot{h}(\tau)| d\tau = 2h(t^*) \tag{A18}$$

where t^* is the zero crossing time for $h(\tau)$, i.e., is the solution of

$$\dot{h}(\tau) = \frac{1}{r_1 - r_2} (-r_2 e^{-r_2 \tau} + r_1 e^{-r_1 \tau}) = 0 \tag{A19}$$

given by

$$t^* = \frac{1}{r_1 - r_2} \log \left(\frac{r_1}{r_2} \right) \tag{A20}$$

Then, the substitution of the expression (A11) for $h(t)$ in (A18) yields

$$\gamma_v(t, \alpha) \leq \lim_{t \to \infty} \gamma_v(t, \alpha) = \frac{2}{r_1 - r_2} e^{-r_2 t^*} (1 - e^{-(r_1 - r_2)t^*})$$

$$\leq \frac{2}{r_1 - r_2} (1 - e^{-(r_1 - r_2)t^*}) \tag{A21}$$

and, taking into account (A20)

$$\gamma_v(t, \alpha) \leq \lim_{t \to \infty} \gamma_v(t, \alpha) \leq \frac{2}{r_1} \tag{A22}$$

Finally, to fulfill condition (A4c), the problem is that of providing a structure to $L(\alpha)$ in order to have (after the substitution of $P(\alpha)$ with α, as required by (A14))

$$\frac{2}{r_1} = \frac{4}{L(\alpha) + \sqrt{L^2(\alpha) - 4\alpha}} \leq \frac{1}{\alpha} \tag{A23}$$

which implies

$$L(\alpha) = 2\alpha + \frac{1}{2}; \qquad \alpha > \frac{1}{4} \tag{A24}$$

Thus, the structure of $L(\alpha)$ and $P(\alpha)$ allowing the fulfillment of the desired properties described by (A2), (A4a), (A4b), and (A4c) have been completely specified. Note that, since the upper bounds in (A4b) and (A4c) vanish for increasing α, and $\gamma_p(t, \alpha)$, $\gamma_v(t, \alpha)$ are increasing time-functions, it is apparent that also (A4d) holds true.

It is, however, possible to proceed further on by providing a useful upper bound (minimal in some sense) for $\|(2\alpha = 1/2)\dot{y} + \alpha y\|_r$. To this aim and by applying the same reasoning line as before, we get

$$\left\|\left(2\alpha + \frac{1}{2}\right)\dot{y} + \alpha y\right\|_t \le \left(\int_0^t \left|\left(2\alpha + \frac{1}{2}\right)\dot{h}(\tau) + \alpha h(\tau)\right| d\tau\right)\|u\|_t \qquad \text{(A25)}$$

In order to properly handle inequality (A25) (i.e., properly compute the right-hand side integral), note that, with the specified choice of $L(\alpha)$ and $P(\alpha)$ the roots r_1 and r_2 turn out to be

$$r_1 = 2\alpha; \qquad r_2 = \frac{1}{2} \qquad \text{(A26)}$$

Actually, by substituting the structure of $h(\tau)$ and $\dot{h}(\tau)$ in (A25) and taking into account (A26) it is not difficult to see that

$$\left\|\left(2\alpha + \frac{1}{2}\right)\dot{y} + \alpha y\right\|_t \le g(t, \alpha)\|u\|_t \qquad \text{(A27)}$$

where a little algebra shows that

$$g(t, \alpha) \triangleq \int_0^t \left|\left(2\alpha + \frac{1}{2}\right)\dot{h}(\tau) + \alpha h(\tau)\right| d\tau$$

$$= \frac{1}{2(4\alpha - 1)}\int_0^t \left|16\alpha^2 e^{-2\alpha\tau} - e^{-\tau/2}\right| d\tau \qquad \text{(A28)}$$

nonnegative, increasing and such that $g(0, \alpha) = 0$. Moreover,

$$g(t, \alpha) = \frac{1}{2(4\alpha - 1)} \left[2(4\alpha - 1) + 2(e^{-t/2} - 4\alpha e^{-2\alpha t})\right] \qquad t \le \bar{t} \quad \text{(A29a)}$$

whereas

$$g(t, \alpha) = \frac{1}{2(4\alpha - 1)} \left[2(4\alpha - 1) + 4(e^{-\bar{t}/2} - 4\alpha e^{-2\alpha\bar{t}})\right.$$
$$\left. - 2(e^{-t/2} - 4\alpha e^{-2\alpha t})\right] \qquad t \ge \bar{t} \qquad \text{(A29b)}$$

being $\bar{t}$ the root of

$$16\alpha^2 e^{-2\alpha\tau} - e^{-\tau/2} = 0 \qquad \text{(A30)}$$

i.e.,

$$\bar{t} = \frac{4}{4\alpha - 1} \log(4\alpha) \qquad \text{(A30bis)}$$

Actually, since $g(t, \alpha)$ is an increasing-time function, we get

$$g(t, \alpha) \leq \lim_{t \to \infty} g(t, \alpha) = \frac{1}{2(4\alpha - 1)} [2(4\alpha - 1) + 4(e^{-\bar{t}/2} - 4\alpha e^{-2\alpha \bar{t}})]$$

$$= 1 + \frac{1}{2(4\alpha - 1)} [e^{-\bar{t}/2} - 4\alpha e^{-2\alpha \bar{t}}] \qquad \text{(A31)}$$

Equation (A31) can be rewritten as

$$\lim_{t \to \infty} g(t, \alpha) = 1 + \frac{2}{4\alpha - 1} e^{-\bar{t}/2} [1 - 4\alpha e^{-(2\alpha - 1/2)\bar{t}}]$$

$$< 1 + \frac{2}{4\alpha - 1} [1 - 4\alpha e^{-(2\alpha - 1/2)\bar{t}}] \qquad \text{(A32)}$$

and the substitution of (A30bis) in (A32) directly yields

$$\lim_{t \to \infty} g(t, \alpha) < 1 + \frac{1}{2\alpha},$$

thus, providing an upper bound for $g(t, \alpha)$. Moreover, by reconsidering the structure of $g(t, \alpha)$ given by (A29a) and (A29b), it is not difficult to see that

$$\lim_{\alpha \to \infty} g(t, \alpha) = 1 \qquad \forall t > 0 \qquad \text{(A33)}$$

Note that condition (A33) holds only for all positive-time instants, since for increasing α the structure of $g(t, \alpha)$ approaches the shape of the unity step. On the other hand, the condition $g(0, \alpha) = 0$ always holds true.

Summing up and trivially extending the above results to decoupled MIMO systems, the following property can be stated

Property One—for a MIMO time-invariant decoupled system described by the vectorial differential equation

$$\ddot{z} + \left(2\alpha + \frac{1}{2}\right)\dot{z} + \alpha z = U \qquad \text{(A34)}$$

being $z \in \mathcal{R}^n$, $U \in \mathcal{R}^n$, and subject to null initial conditions, the following hold true

$$\begin{cases} \|z\|_t \leq \gamma_p(t, \alpha)\|U\|_t \\[2mm] \|\dot{z}\|_t \leq \gamma_v(t, \alpha)\|U\|_t \\[2mm] \left\|\left(2\alpha + \dfrac{1}{2}\right)\dot{z} + \alpha z\right\|_t \leq g(t, \alpha)\|U\|_t \end{cases} \qquad (A35)$$

where $\gamma_p(t, \alpha)$, $\gamma_v(t, \alpha)$, and $g(t, \alpha)$ are positive increasing-time functions, such that

$$\begin{cases} \gamma_p(0, \alpha) = \gamma_v(0, \alpha) = g(0, \alpha) = 0 \qquad \forall \alpha \\[3mm] \lim_{t \to \infty} \{\gamma_p(t, \alpha)\} \leq \dfrac{1}{\alpha} \\[3mm] \lim_{t \to \infty} \{\gamma_v(t, \alpha)\} \leq \dfrac{1}{\alpha} \\[3mm] \lim_{t \to \infty} \{g(t, \alpha)\} \leq 1 + \dfrac{1}{2\alpha} \\[3mm] \lim_{\alpha \to \infty} \gamma_p(t, \alpha) = \lim_{\alpha \to \infty} \gamma_v(t, \alpha) = 0 \qquad \forall t \leq 0 \\[3mm] \lim_{\alpha \to \infty} g(t, \alpha) = 1 \qquad\qquad\qquad \forall t > 0 \end{cases}$$

In (A33) the following definition of vector norm has been used:

$$\|w\|_t \triangleq \sup_{0 \leq \tau \leq t} \|w(\tau)\| \qquad (A37)$$

where $\|w(\tau)\|$ is the usual euclidean norm of vector $w(\tau) \in \mathfrak{R}^n$.

REFERENCES

1. S. Arimoto, S. Kawamura and F. Miyasaki, "Bettering Operation of Robot by Learning," *Journal of Robotic Systems*, Vol. 1, No. 2, 1984, pp. 123–140.
2. S. Arimoto, S. Kawamura and F. Miyasaki, "Can Mechanical Robots Learn by Themselves?," *Second International Symposium in Robotics Research*, MIT Press, Cambridge, MA, 1985, pp. 127–134.
3. S. Arimoto, S. Kawamura and S. Tamaki, "Learning Control Theory for Dynamical Systems," *Proceedings of the 24th. IEEE Conference on Decision and Control*, Ft. Lauderdale, FL, December 1985.
4. J.J. Craig, "Adaptive Control of Manipulators Through Repeated Trials," *Proceedings 1984 American Control Conference*, San Diego, CA, June 1984, pp. 1566–1573.

5. J. Hauser, "Learning Control for a class of Nonlinear Systems," *Proceedings of the 26th. IEEE Conference on Decision and Control,* Los Angeles, CA, December 1987, pp. 859–860.

6. G. Casalino and G. Bartolini, "A Learning Procedure for the Control of Movements of Robotic Manipulators," *Proceedings IASTED Symposium on Robotics and Automation,* Amsterdam, The Netherlands, June 1984.

7. G. Casalino and G. Bartolini, "A Learning Approach to the Control of Robotic Manipulators," *Proceedings National Meeting of the Italian Society for Automation* (ANIPLA), Genova, Italy, December 1985.

8. G. Casalino and L. Gambardella, "Learning of Movements in Robotic Manipulators," *Proceedings 1986 IEEE Conf. on Robotics and Automation,* San Francisco, CA, April 1986.

9. P. Bondi, G. Casalino and L. Gambardella, "On the Iterative Learning Control Theory for Robotic Manipulators," *IEEE Journal of Robotics and Automation,* Vol. 4, No. 1, February 1988, pp. 14–22.

10. M. Kawato, M. Isobe, Y. Maeda, R. Suzuky: "Coordinate Transformation and Learning Control for Visually-Guided Voluntary Movements with Iteration: a Newton-like Method in Function Space," *Biological Cybernetics,* No. 59, 1988, pp. 161–177.

11. R. Ortega and M.W. Spong, "Adaptive Motion Control of Rigid Robots: A Tutorial," *Proceedings of the 27th. IEEE Conference on Decision and Control,* Austin, TX, December 1988.

12. J.J. Craig, *Adaptive Control of Mechanical Manipulators,* Addison Wesley, MA, 1988.

13. A. Kantorovich, *Functional Analysis,* Pergamon, NY, 1982.

3

Vision-Based Robotic Assembly System

Z. Bien
Dept. of E. E., KAIST, Kusong-dong, Yusong-ku, Taejon, Korea

I.H. Suh
Dept. of E. E., Hanyang Univ., Haengdang-dong, Seongdong-ku, Seoul, Korea

S.-R. Oh
Control Systems Lab., KIST, 39-1, Haweolkok-dong, Songbuk-ku, Seoul, Korea

B.-J. You
Robot Division, R & D Center, TURBO-TEK Co., Ltd., 1-2, 1-ga, Wonhyo-ro, Yongsan-ku, Seoul, Korea

1. INTRODUCTION

Automation machines for assemblying microelectronic components, such as semi-conductor die-bonding machines, have become rather complex systems often equipped with mechanical manipulators and sophisticated vision sensors for reliable operation and system flexibility. Physically, these assembly machines consist of a mechanical body with actuators and an electronic control system. Functionally, on the other hand, design of such machines requires integration of various functional units performing visual pattern recognition, guidance of components, control for mechanical manipulators, and sequential control of mechanical fixturing devices and material transport [1, 2].

This chapter describes the design of a multiprocessor-based automatic machine for semiconductor die bonding. The mechanical structure of the machine is configured in consideration of desired operational efficiency of existing machines [3], while the electronic control system is designed to be a master–slave type consisting of several subsystems, each of which employs a 16 bit microprocessor MC68000. A special feature of the control system is that, in comparison with existing machines, manual operation is minimized by introducing automatic measurement and correction of the size and orientation of each component.

The structured supervisory controller, working as a real-time operating system for the multiprocessors with multitask control, effectively coordinates all sensor signals and actuator functions. Various image-processing algorithms, such as projection technique and Hough transform, are incorporated in the control system for fast operation. In particular, a new pattern recognition method is presented for locating objects based on a novel concept called incremental-circle transform. The incremental-circle transform, which maps boundaries of an object onto a circle, effectively represents the shape of the boundaries detected in binary images of the object. Further, an intelligent display system and servo controllers to drive the high-speed stepping motors and DC motors of the manipulators are included in the control system.

2. OVERALL SYSTEM CONFIGURERATION

The outlook of the developed semiconductor die-bonding machine is shown in Figure 3.1. The mechanical body of this assembly machine consists of the following three modules: (1) bonding-head module, (2) wafer-feeding module, and (3) lead-frame feeding module.

The bonding-head module is driven by four actuators, two step motors, and two DC motors. It is utilized to transfer each die from a wafer to an array of lead frames. The plunge-up unit, actuated by a step motor, is also included in this module to push up each die, so that the bonding head picks up a die easily and reliably. The wafer-feeding module consists of a camera with a microscope for inspection and measurement of the posture of a transferred die, an XY-table driven by two step motors for wafer feedings, a jig holder actuated by a step motor, and a wafer loader and unloader. Finally, the lead-frame feeding module is actuated by five step motors, two DC motors, and five pneumatic cylinders. The lead frame, stacked up initially in the frame loader, is fed by the frame feeder and two clampers. When it is positioned at a bonding point, it is fastened by a window clamper. After a die has been bonded, the lead frame is unloaded and stacked in the frame stacker.

The overall control system is organized essentially as a concurrently multifunctional system realized with multiple processors. The control system is designed to be a master–slave type multiprocessor system in which one processor plays the role of the master and the others are the slaves. The processors are tightly coupled with heavy processor interaction. As shown in Figure 3.2, each slave that is a functionally dedicated system in software communicates with each other via a time-shared common bus structure with a hard-wired bus arbitration scheme [4].

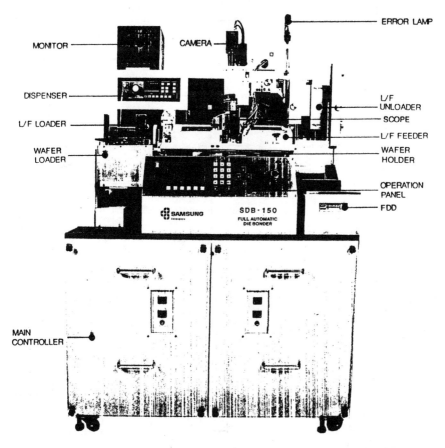

ERROR LAMP

MONITOR

CAMERA

DISPENSER

L/F
UNLOADER

SCOPE

L/F LOADER

L/F FEEDER

WAFER
LOADER

WAFER
HOLDER

OPERATION
PANEL

FDD

SAMSUNG

SDB-150

FULL AUTOMATIC
DIE BONDER

MAIN
CONTROLLER

FIGURE 3.1.   Laboratory Developed Semiconductor Die-Bonder Machine

The intercommunication is carried out by means of *semaphore* messages passed to a shared memory via the parallel bus architecture global bus (**GBUS**). In the specific control system in Figure 3.2, there are three 16 bit microprocessors MC68000 connected to **GBUS** as follows:

1. the system master, called supervisory controller (**SUPC**),
2. the visual pattern recognition unit (**VPRU**),
3. the display unit (**DISU**).

In the system, servo motors and actuators are controlled mainly by **SUPC**. Since the servo control mechanism is designed in a conventional way, emphasis will be given on describing the supervisory controller and the vision subsystem.

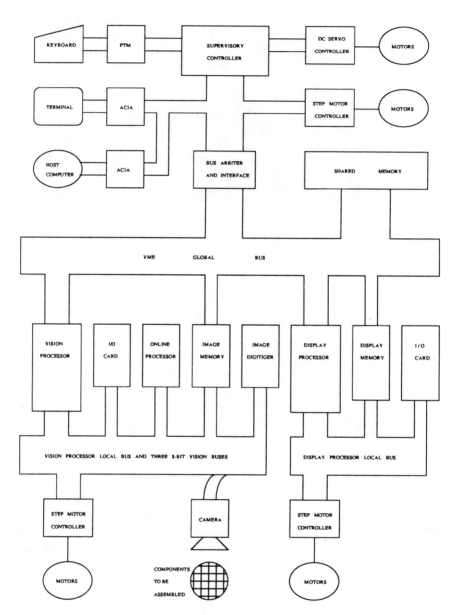

FIGURE 3.2.   Overall System Configuration

3. STRUCTURED SUPERVISORY CONTROLLER
WITH DIAGNOSIS CAPABILITY

To be a satisfactory system, the supervisory controller (**SUPC**) must be capable of effectively coordinating tasks assigned to slaves. It must also provide operators with user-friendly person–machine interfacing features so that both operational mistakes and the system set-up time can be minimized. Furthermore, the design required that self-diagnosis functions be included in the supervisory controller so that the assembly machine can be easily taken care of in case of faults and malfunctions. In order for **SUPC** to accommodate these requirements, the following six function states are introduced:

1. Idle State (IS),
2. Parameter State (PS),
3. Adjust State (ADS),
4. Auto-Assembly state (AAS),
5. Diagnosis State (DS),
6. Emergency State (ES).

In the *Idle State,* **SUPC** receives operational commands while letting **DISU** display the main menu after initializing software procedures and hardwares, such as the timer interrupt handler and motors/actuators. In the *Parameter State,* various parameter values can be input and modified to inform the control system about component characteristics. For the die-bonding machine, such parameters as sizes of wafer, die and ink-dot, width/length of lead frame, and number of lead frames are included. Also, parameter values for the visual image processing are, for example, threshold value, image intensity, and window size. In the *Adjust State,* **SUPC** lets each slave precisely tune the motions of actuators interfaced to each slave and helps human operators to heuristically choose data that may be required for reliable visual-signal processing. In the *Auto-Assembly State,* **SUPC** not only performs its own tasks, but also correctly assigns necessary tasks to each slave by monitoring the degree of completion of the previously given tasks to each slave. In the *Diagnosis State,* **SUPC** first examines the hardware modules linked to the microprocessor and then makes **DISU** display the current on-off states of all the sensor/actuator signals, so that the human operator easily monitors malfunctions of the sensors. Finally, in the *Emergency State,* **SUPC** shuts down all the motors/actuators and finds out and displays which limit switches are turned on outside the operating ranges.

To effectively handle these functional states, the supervisor **SUPC** is designed as a multilayered structure type similar to the **UNIX** operating system [5], as shown in Figure 3.3. That is, in the core of the **SUPC** resides virtually all the system hardware, such as AC motors, DC motors, step motors, limit-switch sensors, pneumatic solenoids, and D/A and A/D converters. Thus, **SUPC** can

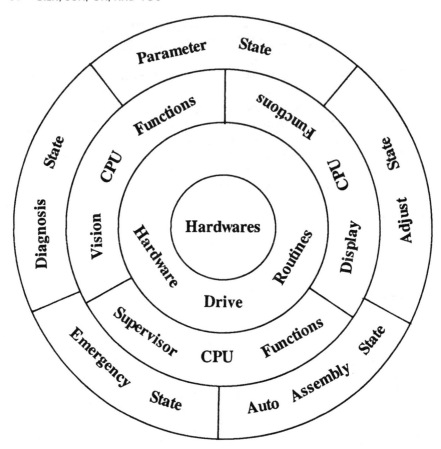

FIGURE 3.3. Layer Structure of Supervisory Control

indirectly control and monitor all the hardware with the help of the slaves. In the next layer, there are software procedures for driving the system hardware corresponding to the kernel of the **UNIX.** The outer layer next to the kernel is called the *Function Layer,* and includes primitive actions of **SUPC** as well as those of each slave. There are about 150 primitive actions for the semiconductor die bonding (70 for supervisor and 40 for each slave). The outermost layer consists of the six functional states, described previously. Because of its flexibility and modularity, the supervisor of such a multilayered structure can be easily modified and expanded when other system functions must be added.

The supervisor tasks are divided into foreground and background tasks. The background task is performed periodically according to a ten msec timer interrupt request. The supervisor checks each keyboard input command and then puts it into one of the three command queues according to its task priority. **SUPC** also checks the error flag, and if the error flag is set, **SUPC** resets the system. Otherwise, **SUPC** reads and monitors the status of each slave to determine if *SUPC* can assign the new task to each slave.

The foreground, on the other hand, are the main tasks performed in each state. First, **SUPC** chooses a task to perform from the command queue with the highest priority. If there is no command in the queue, then **SUPC** chooses a task in the second priority command queue. If there is no command in the second priority queue, **SUPC** finally searches for a task in the lowest priority command queue. **SUPC** then assigns the task, if any, to the predesignated slave via the shared memory. In the system, *system reset* and *emergency stop* are included in the highest priority group, and vision-related tasks, such as *binary image display* and *display of bonded die number,* are included in the next priority group. Almost all other jobs belong to the lowest priority group. In this way, the supervisor (**SUPC**) can immediately serve for the real time command request while the system works for the current user's command.

Intercommunication between the master and the slaves is done through the shared memory. To ensure that the global bus (**GBUS**) is used by only one processor at a time, a hard-wired bus arbitration is employed, while a semaphore message as in Figure 3.4 is utilized to avoid the overwriting. Here TAS (test and set) instruction of the MC68000 CPU can be used to see if a common memory location is now being occupied by one of other processors.

In the design of the **SUPC** for the die-bonding machine, the system faults are grouped into the following three levels:

- (level 0)—faults detected during the automatic-diagnosis state to check the processor-linked hardwares after power-on,
- (level 1)—faults occurred during on-line operation, such as dropping the parts,
- (level 2)—faults detected by the stroke-limit sensors.

These system faults should be detected and isolated in such a way that mechanical damage to the system, if any, is minimum, and the system can be recovered as fast as possible to reduce the down-time. To do this, **SUPC** is endowed with the capability of self diagnosing each processor module including the central processing unit (CPU), memory, bus, and timer, and also letting the human operator easily monitor malfunctions of the system by displaying the current on-off states of all the sensor/actuator signals.

For the self-diagnosis of MC68000 CPU in each processor module, registers and internal data bus in the MC68000 CPU are first tested by making the least significant bit (LSB) set to zero and others set to one and by moving the data from one register to another register sequentially via internal CPU bus. Then the transfered data is identified to be correct. Secondly, ALU functions are tested by performing each ALU related instruction and by comparing the result with the

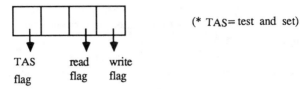

(* TAS= test and set)

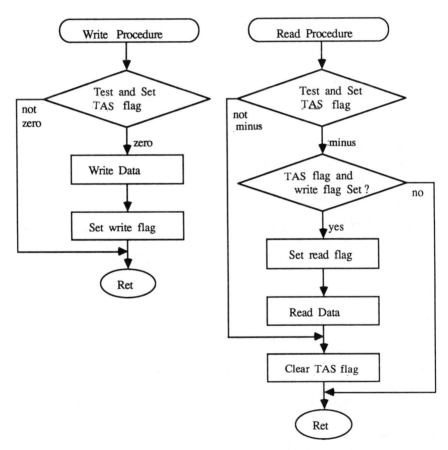

FIGURE 3.4.   Communication Protocol for Mutual Exclusion

precomputed value. Thirdly, addressing modes are tested by reading and identifying the data in a known memory address with each addressing mode.

It is noted that almost all CPUs have been made to detect and process exceptional events, such as bus-trap error, address-trap error, illegal-instruction error, and devide-by-zero error. When such an event occurs, CPU usually stops or holds all of its functions. Thus, the CPU-based controller cannot control the system any more regardless of the machine state, which may cause the machine to be out of order. To overcome such difficulties, the exception processing routine shown Figure 3.5 is designed and included. The routine is then performed by changing the exception vector address to the starting address of the routine when the exceptional event occurs.

The self-diagnosis method of the memory and bus is very similar to that of internal registers and bus in the CPU. Since periodic timer interrupt is the core in the background task of the **SUPC,** its function should be monitored by the operator when diagnosing. This is done by sending a predefined message to the **DISU** whenever the interrupt occurs. Figure 3.6 shows the self-diagnosis results.

For the diagnosis of input/output and other modules, the current on/off states of all the sensors and relays are displayed. Each function of the assembly machine is performed one-by-one. These features let the operator easily monitor malfunctions of the sensors, relays, and the actuators. Figure 3.7 shows the sensor-status display on the monitor.

4.   VISUAL PATTERN RECOGNITION SYSTEM

Before the actual automatic assembly task takes place in the machine, the position and orientation of the components should be measured and determined in real time, and at the same time, the quality of each component must be inspected to find out whether the component is a good die or a bad die. For this, a visual-pattern recognition system is designed that is capable of fast processing. The structure of the developed vision system is shown in Figure 3.8.

To transfer image data at video rate (6.9 MHz) between image processors, three eight-bit vision buses in VME I/O-bus are incorporated. Especially, to reduce the time of transferring image data from the camera to the pattern-recognition processor, a new type of frame grabber is designed, which is able to input or output the image data at video rate. Also, to reduce the on-line processing time for quality inspection and calculation of position for each component, a hard-wired on-line processing unit is proposed. As the vision sensor, a CCD camera that uses external vido synchronization signals is adopted, and a microscope is attached for recognition of small components, such as integrated circuit chips. The spatial resolution of the vision system is 256 × 256, and the brightness resolution is 256.

Included also is pattern-recognition software that is relatively fast and insensi-

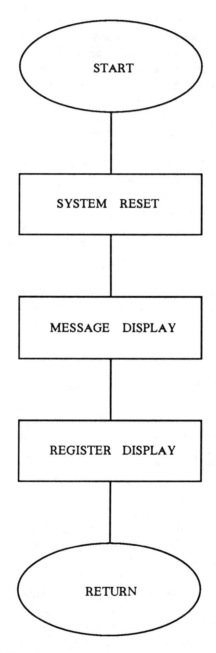

FIGURE 3.5.   Exception Processing Routine

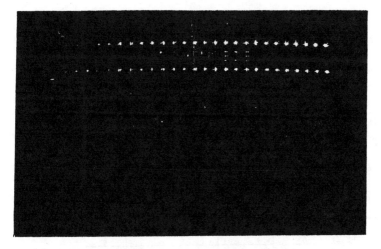

FIGURE 3.6.   Self-Diagnosis Results

tive to noise by employing the Hough transform technique [6] with a simple edge detection, the projection method, and modified edge-preserving smoothing algorithms [7]. Especially, a fast and systematic approach for determining the posture of randomly placed collision-free objects is proposed, based on a new concept of incremental-circle transform. The incremental-circle transform represents the

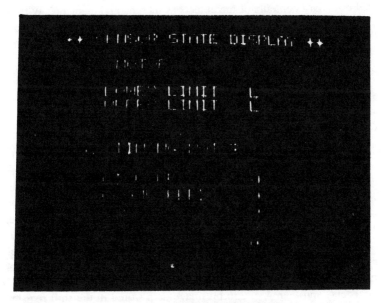

FIGURE 3.7.   Sensor Status Display

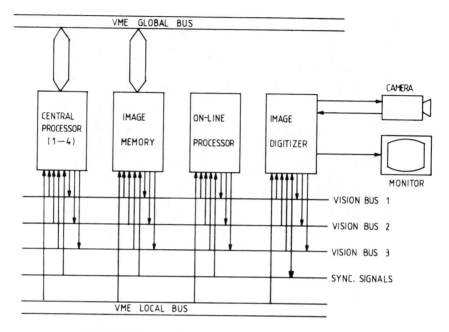

FIGURE 3.8.   Structure of Designed Vision System

detected boundary contours of the objects as incremental elements. With the aid of similarity and the line integral, the transform quickly finds the orientation of each object independent of the position and starting point of the boundary coding.

4.1.   Hardware Structure of Vision System

4.1.1.   *Video-Rate Frame Grabber*

To reduce the image-data transfer time and to endow the central processor (CPU) with real time pattern-recognition capability, it is required that the image memory is used for the system memory to which the CPU can access whenever it desires. Consequently, memory components with two channels for data transfer are needed and so a commercially available dual-ported dynamic random access memory (DRAM), called TMS4461 [8], is adopted. The most interesting feature of the TMS4461 is that 4×256 bit shift-registers with fast access time are connected to the conventional $64K \times 4$ bit dynamic RAM in a parallel way. Thus, in TMS4461, the digitized visual-image data are first stored in the serially connected 4×256 bit shift-registers, and then 4×256 bit image data stored in the shift registers are transferred to the dynamic RAM within 120 nsec. It is remarked that only 120 nsec

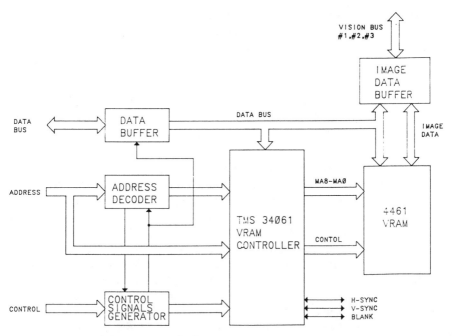

FIGURE 3.9.   Block Diagram of Image Memory

is required to use the dynamic memory to store a line data per every 63.5 microsecond. This enables the CPU to use the video memory for other purposes. There is some design difficulty, however, in using TMS4461 due to different types of addressings required by CPU, refreshment, and line generator. This latter difficulty is resolved by employing a TMS 34061 VRAM controller, which can arbitrate the addressing requirements by refreshment, CPU, and line generator. The block diagram of the designed-image memory is shown in Figure 3.9.

4.1.2.   On-Line Preprocessing Unit.

The visual pattern recognition for the component assembly can be divided into the off-line processing and the on-line processing. In the case of the off-line processing, process parameters including the threshold value for binarization and the number of image pixels corresponding to the minimal movement of the XY-table should be determined with high precision.

On the other hand, in the case of on-line processing, the quality inspection and calculations of the positional and orientational deviations from the pre-specified pick-up position are performed for each object. Thus, some processing algorithms for the on-line function is realized by hardware circuits, called as on-line processing unit, for fast data processing.

The on-line processing unit is composed of digital comparators to make the grey-level image into the binary image by means of a threshold value, window and line counters, and the interrupt requester. Here, the window counters play the role of counting the number of black pixels in a line, where the number of black pixels in the windows will be used in quality inspection and in finding out the positional deviations. Thus, the hard-wired window counters may reduce the on-line processing time. The hardware block diagram of the on-line processing unit is shown in Figure 3.10.

4.2.   Pattern-Recognition Algorithms

The vision system is designed for automatic assembly machines of small simple-looking electronic components, such as rectangular-shaped integrated-circuit chips with quality marks as in a die-bonding machine [3, 9] or the similar. The die-bonding machine transfers and attaches good die in a wafer one at a time to a sequence of lead frames while conducting automatic correction of positional and orientation error. In a modern electronic wrist-watch maker, one may find a kind of die attachment machine that is used for assembling integrated-circuit watch chips onto a printed circuit board for an electronic watch. In this latter case, the machine is fed with rectangular components contained in a regular array, called waffle pack.

For recognition of these types of components, several algorithms are developed as follows.

4.2.1.   Thresholding Algorithm.

In the case of the semiconductor die-bonding machine, the rectangular shape of semiconductor dies, kerfs, and metal-base constitute the wafer. The grey-level image of the wafer should be binarized in such a way that metal-base and semiconductors become white image and kerfs become black image. However, brightness of the semiconductors and that of kerfs are slightly different, while the brightness of the semiconductors and that of the metal-base is quite different. Thus, binarization by a single thresholding value is doomed to fail. In addition to this difficulty, the histogram of the image, in this case, is quite noisy.

To solve these difficulties, a thresholding method using the concept of the seperability function in [10] is here employed, and the result is found to be more robust than other methods. The method is first applied to the image to find out the thresholding value that discriminates the metal-base and the rest. Then the whole image is modified such that the images whose grey level is greater than the threshold value are temporarily eliminated, and then the thresholding method is reapplied to the modified image to find out another thresholding value to discriminate silicons and kerfs. The experimental result on the binarization by the proposed thresholding method is shown in Figure 3.11.

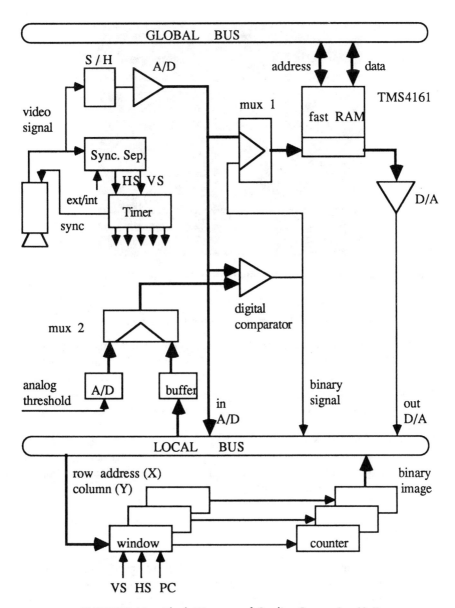

FIGURE 3.10.   Block Diagram of On-line Processing Unit

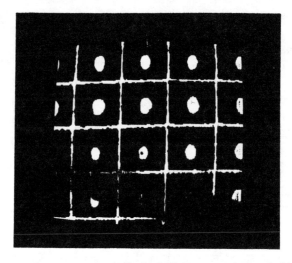

FIGURE 3.11.   Binary Image Results from the Proposed Thresholding Method

4.2.2.   Measurement of Posture for a Tilted Component.

In the case of the semiconductor die-bonding machine, the die may be put on the mylor with a tilted posture. Typically most commercial die-bonding machines allow the working range of the titled angle of a die to be within ±5 degrees. In this section, an efficient algorithm [11] is described in the following to determine the posture of the slightly tilted components under the assumptions:

1. Each component is rectangular-shaped.
2. If the size of a component is $S_x \times S_y$ and the positional deviations of the component relative to the origin of image plane are denoted as D_x and D_y along x-axis and y-axis, respectively, then,

$$-S_x/2 < D_x < +S_x/2,$$
$$-S_y/2 < D_y < +S_y/2.$$

It is noted that the second assumption is given to specify the position of recognition windows and thus speed up the recognition time. The size of the window depends on the quantity of positional deviation of a component with respect to a reference point, while the position deviation is determined by the accuracy of a feeding mechanism of the machines. Thus, the assumption is reasonable in the sense that the positional deviations of the feeding mechanism is much smaller than the size of the component in high-precision assembly machines, such as the die bonder, the die attachment machine, or an automatic insertion machine of parts.

 4.2.2.1.   Orientation Detection. As depicted in Figure 3.12a, the boundary

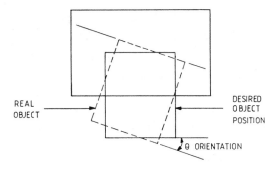

(a) Orientation θ

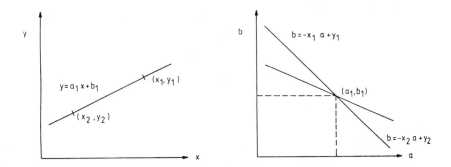

FIGURE 3.12.   Orientation Detection

of a rectangular-shaped component consists of parallel straight lines. Thus, the orientation of the component can be determined by calculating the line equation for one of the boundary-line segments. To reduce the processing time, a recognition window is set up on the upper region of the component and the corresponding line equation is determined via the Hough transform method [6], which will be described shortly. Then, the orientation is calculated from the slope of the line. It is remarked that the least-square-error method [12] is found not suitable for the above task of orientation detection due to its rather high sensitivity to noise.

The Hough transform method finds an equation of a line by converting pixels in the image space into points in the two-dimensional parameter space whose coordinates denote the slope and the y-intercept value of a straight line, respectively. For example, suppose that a line L, in the x-y image space, is given and described by the line equation of $y = a_1 x + b_1$, as shown in Figure 3.12b. The line L is then mapped to the set of lines, $\{b = -x_i a + y_i | (x_i, y_i) \in L\}$, all of which pass through the point, (a_1, b_1) in the a-b parameter space, as shown in Figure 3.12c. In the case when the line L in the image space is contaminated somewhat by noise, the Hough transform method renders the slope and the y-intercept of the equation of the line by identifying the most clustered point through which most of the lines pass in the parameter space.

To use the Hough transform method, there are presteps for repetitive operations. The first step is to specify the range of the slope, a, and that of the y-intercept, b, of the boundary. For this, let

$$|a| < A_m \quad \text{with} \quad 0 < A_m < 1$$
$$B_L < b < B_H \quad \text{with} \quad -S_y/2 < B_L < B_H < +S_y/2$$

The second step is to quantize the slope and the y-intercept, so that the accumulator arrays, $A(a, b)$ for parameter space can be constructed. In this step, the resolution of orientation is determined. The final step is to set the recognition window for fast processing. Let the range of the slope by $|\theta| < \theta_m$. Then, the size of the recognition window, $W_x \times W_y$ becomes

$$W_x = 2 \cdot S_x,$$
$$W_y = 2 \cdot (S_y/2 + S_x \cdot (\tan \theta_m)/2)$$

After these presteps, the following algorithm, which will be called the windowed Hough transform, is executed repeatedly in a coordinate frame with its origin at the center of the recognition window.

- Step (1) Input binary images.
- Step (2) Clear all the accumulator array, $A(a, b)$ in the parameter space.
- Step (3) Find candidates for upper edges of the component. A candidate for the edges is found at $(x, y + 1)$ such that

$$p(x, y) = 1 \text{ (background) and } p(x, y + 1) = 0 \text{ (components)}$$

where

$p(x, y)$: pixel value at (x, y) in image plane
x: column address in image plane
y: row address in image plane

- Step (4) Transform the candidates for the edges in the image plane into lines in the parameter space. Increase the value of $A(a, b)$ by one whenever a line $b = -ax + y$ passes through a point (a, b) in the parameter space.
- Step (5) Find accumulator array, $A(a_m, b_m)$ which contains the maximum value in $A(a, b)$. Then determine the orientation, θ of the component.

$$\theta = \arctan(a_m)$$

4.2.2.2.   Position Detection. Once the orientation of a component is corrected to be in line with the image coordinate frame, the center of a component can now be found from the positions of upper and left boundary of the component. To detect the boundaries, the projection method is applied to windowed binary images, as shown Figure 3.13. In Figure 3.13, the window depicted as a solid line denotes the desired position at which the component is placed. The window depicted as a dotted line represents the real component and the recognition window is shown by the dot-dashed line.

Since projection implies integration of the intensity of all the pixels in a specified direction, the effect of noise due to quantization and incomplete threshold is eliminated. It is remarked that when other coding methods of binary images such as chain coding or run-length coding techniques are used, two

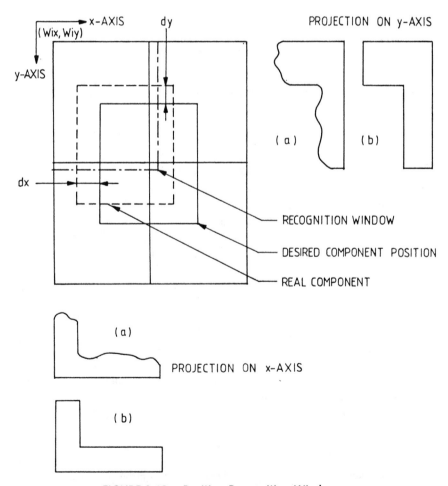

FIGURE 3.13.   Position Recognition Window

adjacent components may be misjudged as one component if any noise exists between the two components. Also, processing time of the projection method is reduced since two-dimensional images are converted into one-dimensional array data. Further, any markings on the component can be easily distinguished from the boundary of the component adopting the projection. The proposed algorithm is as follows:

- Step (1) Input binary images.
- Step (2) Project binary images in the recognition window on x- or y-axis of image plane.

$$P_x(k) = \sum_{i=il}^{im} p(k, i), \quad kl \leq k \leq km$$

$$P_y(k) = \sum_{i=in}^{io} P(i, k), \quad kn \leq k \leq ko$$

$$P(k) = P_x(k) \text{ or } P_y(k)$$

where

$$0 \leq il(kl) < im(km) < 256$$
$$0 \leq in(kn) < io(ko) < 256$$

The projections are modified based on algorithms in [7].
- Step (3) Find differences of the projections (i.e.,

$$DP(k) = P(k + 1) - P(k), \quad kl(kn) \leq k \leq km(ko)).$$

- Step (4) Find a set, T_{min} of indexes which have t minimum values in $DP(k)$ in Step (3)

$$T_{min} = \{k_1, k_2, \ldots, k_t\}, \quad t_m < t < t_{max}$$

- Step (5) If $P(k) > S_x$ (or S_y) $- P_e$, $0 < P_e < P_{max}$, $k \in T_{min}$, then determine a point $(k + 1)$ as a boundary of the component and go to Step (6), else remove k from the set, T_{min} and return to Step (4).
- Step (6) Determine the boundary, B_k.

$$B_k = W_{ix} \text{ (or } W_{iy}) + k$$

It is remarked that when the size of any markings on the component is varying, a fixed threshold in Step (5) does not distinguish the boundary from the

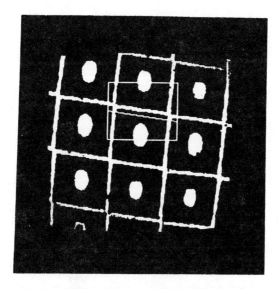

FIGURE 3.14. Experimental Result of Orientation Measurement

markings on the component, and for this case the dynamic thresholding method of projections in Step (3) is adopted.

Experimental results on the proposed method is shown in Figure 3.14. In the figure, the line depicted within the window represents the detected slope of a die using the proposed method, that is windowed Hough transform. Further, in Figure 3.15, we compare the processing time of the proposed method with the SRI method [13] and the conventional Hough transform method without a window. As revealed in the figure, the processing time of the proposed method increases as the recognition range becomes wider, as expected, the processing time for the enlarged parameter space in the windowed Hough transform would take more time. As a result, it is found that the proposed method shows high-speed recognition performance compared to other methods if the orientation angle, $|\theta|$, is less than 30 degrees.

4.2.3. *Measurement of Posture for Randomly Placed Components.*

In the case of the die attachement machine, the die contained in a waffle pack is placed randomly, and hence the method described in the previous section is not effective in view of the processing time. In this case, to measure the orientation and position of the randomly placed object component, a new pattern-recognition algorithm based on a concept of incremental circle transform [14] is described in this section. In the following, a vector is considered as a column vector if denoted otherwise. Given a vector Z or a matrix A, its transpose is denoted as Z^T or A^T, respectively. Also, acronym ICT shall stand for incremental circle transform.

Let us consider a simple closed curve in Figure 3.16, which represents a

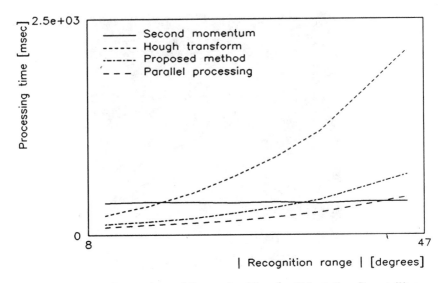

FIGURE 3.15.   Comparison of Processing Time for Orientation Recognition

boundary contour of an object lying on the x-y plane. Further, let the curve, C, be expressed by a parametric vector function [15, 16],

$$\alpha(t) = (x(t), y(t))^T, \ 0 \le t \le L \tag{1}$$

where L denotes the total length of the curve C. An increment of t in $\alpha(t)$ results in counterclockwise-(CCW) directional displacement of the curve. Then, the incremental circle transform is defined as follows [14].

Definition—Incremental Circle Transform

For an arbitrary fixed constant r, let $\Delta_r\alpha(t)$ be given by

$$\Delta_r\alpha(t) = (\Delta_r x(t), \ \Delta_r y(t))^T \tag{2}$$

where

$$\Delta_r x^2(t) + \Delta_r y^2(t) = r^2$$

and there exists $0 \le t \le L$ such that

$$\alpha(t + \Delta t) = \alpha(t) + \Delta_r\alpha(t)$$

We call $\Delta_r\alpha(t)$, $0 \le t \le L$ as the incremental circle transform of $\alpha(t)$, $0 \le t \le L$.

In practice, the incremental elements at a point $\alpha(t)$ on C, $\Delta_r x(t)$ and $\Delta_r y(t)$, are determined by constructing an incremental circle, I_c, with radius r and the center being at the point $\alpha(t)$ on C, and locating in CCW-direction and intersection between I_c and C as shown in Figure 3.16, while $\Delta_r\alpha(t)$, $0 \le t \le L$ can be

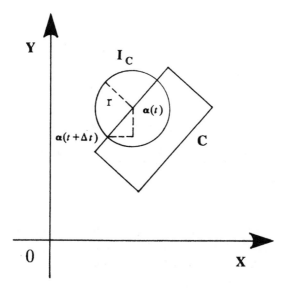

FIGURE 3.16.   Curve and Incremental Circle

obtained by traversing I_c along C continuously in CCW-direction. It is remarked that the definition of the incremental circle transform implicitly assumes that:

1. the boundary contour, C, of an object is piecewise smooth, and
2. $C \cap I_c = O, t \in [0, L]$

Let C_{RT} expressed as $\alpha_{RT}(t)$, $0 \leq t \leq L$, represent a dislocated contour of a curve, C denoted as $\alpha(t)$, $0 \leq t \leq L$. Further, let $\Delta\alpha_{RT}(t)$, $0 \leq t \leq L$, and $\Delta\alpha(t)$, $0 \leq t \leq L$, be ICTs of C_{RT} and C, respectively.

The ICT has several useful properties as follows.

1. Position Invariance:
    The ICTs of translated objects are equivalent to each other.
2. Preservation of Rotational Relationship:
    The rotational relationship between $\alpha(t)$, $0 \leq t \leq L$, and $\alpha_{RT}(t)$, $0 \leq t \leq L$, is the same the rotational relationship between $\Delta\alpha(t)$, $0 \leq t \leq L$, and $\Delta\alpha_{RT}(t)$, $0 \leq t \leq L$.
3. Similarity:
    Let

$$M_t = \int_C \Delta_t\alpha(t) \cdot \Delta_t\alpha^T(t) \, dt,$$

$$T_t = \int_{C_{RT}} \Delta_t\alpha_{RT}(t) \cdot \Delta_t\alpha_{RT}^T(t) \, dt$$

Then,

$$T_I = R \cdot M_I \cdot R^{-1} \tag{3}$$

where

$$R = \begin{bmatrix} \cos\theta & -\sin\theta \\ \sin\theta & \cos\theta \end{bmatrix}.$$

For orientation of an object, relative orientation, and absolute orientation are considered.

1.   Absolute Orientation, θ_a

The θ_a represents the attitude of an object with respect to the base coordinate of the image plane. The θ_a can be obtained from the matrix R, M_1 in (3) which diagonalizes the matrix. As the result, the absolute orientation is determined as follows:

$$\theta_a = \begin{cases} \dfrac{1}{2} \tan^{-1}\left(\dfrac{2b}{c-a} \right) & \text{if } a \neq c \\[2mm] \dfrac{\pi}{4}, -\dfrac{\pi}{4} & \text{if } a = c \text{ and } b \neq 0 \\[2mm] undefined & \text{if } a = b = c = 0 \end{cases}$$

where

$$M_I = \begin{bmatrix} a & b \\ b & c \end{bmatrix}$$

2.   Relative Orientation, θ_r

The θ_r describes the attitude relationship of an object relative to another. The relative orientation is determined in the average sense based on three equalities in (3), independent of positions of objects and starting point for ICT of each contour, as follows:

$$\theta_r = \frac{1}{3} \cdot \sum_{i=1}^{3} \theta_i$$

3.   Position Detection

The position of an object is determined based on the concept of the center of mass in [16] by using only coordinates of boundary contour of the object. Let the boundary contour be expressed by $\alpha(t)$ shown in (1). Then, the position of the object, $(x_c, y_c)^T$, is determined as follows:

$$x_c = \frac{1}{L} \int_L x(t) \, dt$$

$$y_c = \frac{1}{L} \int_L y(t) \, dt$$

The proposed algorithm is applied to a machine developed for die attachment [14]. For the incremental-circle transform, in the previous stage of repetitive assembly, a digitized-incremental circle, shown in Figure 3.17, with radius ten (pixels) is implemented and the incremental-circle transform is conducted using only addition and subtraction. Next, for detection of orientation of the chips, the incremental-circle transform of several titled watch chips, shown in Figure 3.18, is performed. Resulting incremental elements, $\Delta_r x(t)$ and $\Delta_r(t)$ are shown in

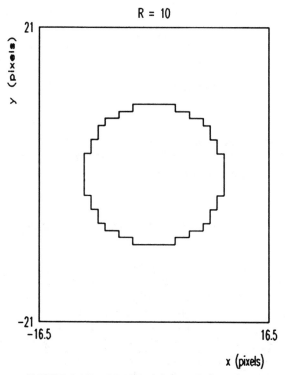

FIGURE 3.17.   Discrete Incremental Circle

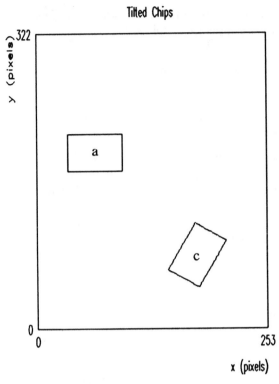

FIGURE 3.18. Tilted Chips

Figure 3.19. Also, values of the matrix M_1 in the Equation (3) and extracted orientations are listed in Figure 3.20. The processing time, about 150 msec, is used up for a 60 (pixels) × 40 (pixels) watch chip.

5. INFORMATION DISPLAY UNIT

As a means of interfacing a person with a machine, a display unit is developed. The display unit **DISU** is composed of a camera interface, a set of double buffer memory (called Vision RAM and Display RAM), and a display processor for displaying in real time (i.e., 63.5 microseconds) raw image, binary image, message and graphic patterns, such as line, circle, and characters.

The image digitizer, as shown in Figure 3.21, is connected to the camera and to a black-and-white monitor. The analog-to-digital converter (ADC) converts analog video signals into 256 × 256 (64K) byte digital grey-level images with a

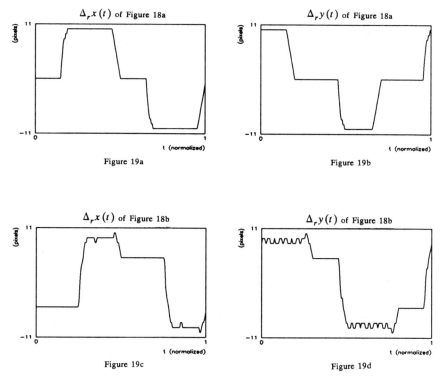

FIGURE 3.19.   Incremental Elements of Objects in Figure 3.18

6.9 MHz sampling rate, while the digital-to-analog converter (DAC), with binary image overlay capability, generates analog video signals from processed digital images. Moreover, synchronization signals generation unit for pixel clock and video synchronizing signals and phase locked loop (PLL) unit for cameras using internal synchronization signals are included.

For graphic-pattern generation, software algorithms for generating line, circle,

		$\int \Delta_r x^2(t)\, dt$	$\int \Delta_r x(t)\Delta_r y(t)\, dt$	$\int \Delta_r y^2(t)\, dt$	θ^o (x-axis)
a	0.0	59.58000	0.00000	39.58000	0.0
b	60.0	43.42735	-8.98544	53.85173	60.05

FIGURE 3.20. M_i and Estimated Orientation

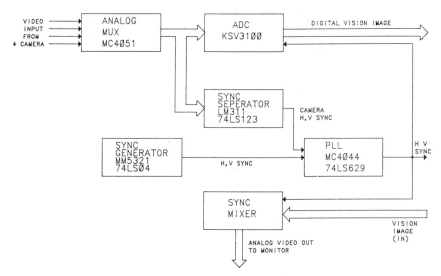

FIGURE 3.21.   Block Diagram of Image Digitizer

and perspective view, as well as character generation, are developed by employing the conventional pattern-generation algorithms [17, 18].

6.   CONCLUDING REMARKS

A design experience for an automatic assembly control system for microelectronic components was presented. The system consists of a structured supervisory controller, a visual inspection/measurement subsystem, an information display subsystem, and a servo subsystem. It is remarkable that the proposed system is flexible and reliable due to modular design and the structured **SUPC** with diagnosis capability. The system minimizes the manual operation via the automatic measurements and corrections of object orientation, position, and component size with the aid of the novel type of the vision system.

The proposed control system was implemented and applied to lab-type semiconductor die-bonding machines and die-attachment machines. It was also successfully tested in real time.

REFERENCES

1.   I. Cibej, B. Solar, T. Bajd, D. Rudel, A.R. Kraj, Z. Babrda, and I. Verdenik, "Robotic Assembly System in Semiconductor Industry," *IEEE Trans. Industrial Electronics,* Vol. IE-34, No. 4, November 1987, pp. 413–416.

2. Z. Bien, M.J. Youn, Y.S. Oh, K.M. Huh, Y.S. Shin, and I.H. Suh, "A Multimicroprocessor-based Control System for Assembly of Electronic Components", Proceedings of LCA '86 IFAC Symposium on Components, Instruments, and Techniques for Low Cost Automation and Applications, Valencia, Spain, Nov. 1986, pp. 155–160.

3. *6300 Die Bonder Manual,* Shinkawa, Japan, 1982.

4. P.H. Enslow, Jr., "Multiprocessor Organization-A Survey," *Computing Surveys,* Vol. 9, No. 1, March 1977.

5. D. Comer, *Operating System Design: The XINU Approach,* Prentice-Hall, NJ, 1984.

6. J. Skalansdy, "On the Hough Technique for Curve Detection," *IEEE Trans. on Computers,* Vol. C-27, 1976, pp. 923–926.

7. M. Nagao, T. Matsuyama, "Edge Preserving Smoothing," *Computer Graphics and Image Processing,* Vol. CGIP-9, 1979, pp. 394–407.

8. *"Mos Memory Databook,"* Texas Instruments, U.S.A., 1989.

9. (Video) *Auto Die Bonder: Model 8030A Manual,* FOTON Automation, U.S.A., 1982.

10. D.H. Ballad and C.M. Brown, *Computer Vision,* Prentice Hall, NJ, 1982.

11. B.-J. You, Y.S. Oh, and Z. Bien, "A Vision System for an Automatic Assembly Machine of Electronic Components," *IEEE Trans. on Industrial Electronics,* Vol. 37, No. 5, Oct. 1990, pp. 349–357.

12. A.M. Mood, F.A. Graybill, D.C. Boes, *Introduction to the Theory of Statistics,* Mcgraw-Hill, NY, 1982, pp. 498–499.

13. G.J. Agin, and R. O. Duda, "SRI Vision Research for Advanced Industrial Automation," Proceedings of Second USA-JAPAN Computer Conference, Tokyo, Japan, Aug. 1975, pp. 113–117.

14. Z. Bien, B.-J. You, H. Lee, S.-R. Oh, and Y. Kim, "A New Pattern Recognition Algorithm with Applications to Real-time Assembly Machine for Two-dimensional Objects", Proceedings of International Conference on Automation, Robotics and Computer Vision, Singapore, 1990, pp. 851–855.

15. C.T. Chen, *Linear System Theory and Design,* CBS College Publishing Co., New York, 1984.

16. M. Hu, "Visual Pattern Recognition by Moment Invariants," *IRE Trans. on Information Theory,* February, 1962, pp. 179–187.

17. B.K.P. Horn, "Circle Generators for Display Devices," *Computer Graphics and Image Processing,* Vol. CGIP-5, 1976, pp. 280–288.

18. J.E. Bresenbaum, "Algorithm for Computer Control of a Digital Plotter," IBM System Journal, Vol. 4, No. 1, 1965, pp. 25–30.

4

Robot Accuracy Issues and Methods of Improvement

Chia P. Day
Alphabet, Inc.
Warren, Ohio

1. INTRODUCTION

Robots have been used in a large number of applications. They are generally flexible, reprogrammable, and repeatable from cycle to cycle. However, for many new industrial applications, just being repeatable will not be enough to satisfy requirements of the job. Robots need to be "accurate" in the sense that the physical location or trajectory reached by the end effector is what the robot has been commanded to.

The importance of robot accuracy has been recognized in early years of robotics development [1, 2, 3, 4], but it takes some time for the necessary technology in algorithm development and application development to realistically and economically overcome robot inaccuracy. Perhaps we have not quite reached a stage when robots can be produced accurately within desired cost, but we certainly have improved a great deal over the years.

1.1. Sensor-Guided Applications

Nowadays, we are beginning to use robots in an increasing number of sensor-guided applications. Sensors, such as vision, are often employed to detect offsets of part location. Offsets are then communicated to robot controllers so that the robot can correct its preprogrammed paths and move to the desired location [5]. The correlation between sensor-detected offset and the actually compensated move must be high. As shown in Figure 4.1, sensor detection is made in sensor-reference frame. Robot movements are commanded using a different reference frame (e.g., tool frame). The task requirements are specified in a goal frame. To transform from one frame to another, several robot parameters, such as link lengths, joint angles, etc.) are used in the transformation matrix. All parameters used in transform calculations are susceptible to error. Factors contributing to robot inaccuracy must be reduced in order to make compensations correct.

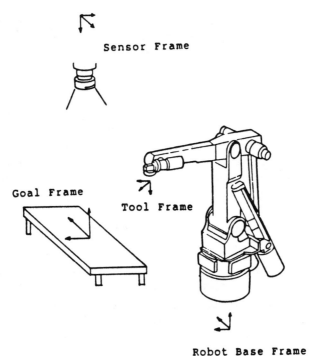

FIGURE 4.1.   Robot Reference Frames

1.2.   Offline Robot Programming

Offline robot programming has been recognized as an important productivity-improvement tool [6]. Robot programming effort can be greatly reduced with the aid of interactive computer graphics. Interactive graphics produce user friendly simulations of robot control commands and robot motions. The end results are robot programs generated from simulations without actually running the robot. This not only reduces the need to tie up a robot for teaching, but also improves programming efficiency, since much of the data on the workpiece and the robot have already existed from computer-aided designs. In order to make an offline programmed robot system truly functional, "accurate" robots are needed. If it were difficult to obtain a truly "accurate" robot, at least a system that ensured conformity of actual robot movements to simulated offline-generated robot movements would be required.

1.3.   Teach-and-Repeat Method

Many industrial robots are still being taught by the teach-and-repeat method, rather than offline programming. Teaching is done by using a teach pendant to jog robot axes, or by a lead-through method where the robot is guided by human force to a desired location. The desired locations of robot movement are then recorded. Trajectories are generated through these recorded points. Again, due to inaccuracies of robot parameters and due to errors in robot controls, the actual playback trajectory may not coincide with the taught locations or trajectory. For some applications, such as painting, slight inaccuracy is tolerable. For other applications, such as precision assembly, inaccuracy cannot be allowed. In this chapter, it is intended to provide readers with an understanding of major factors contributing to robot inaccuracy. Several methods to improve robot accuracy are discussed. With acquaintance of relevant issues, successful applications can be realistically planned and desired productivity can be actually achieved.

2.   ACCURACY DEFINITION

There is not yet a unified defintion of robot accuracy. The International Standards Organization, the Robot Industries Association, and the National Bureau of Standards (now National Institute of Standard and Technology) have been making strides towards definitions of accuracy and devising standard tests to make performance measurements of robot accuracy [7, 8].

For consistency of discussions in this chapter, the following definitions are used.

- Robot Accuracy—A measure of difference between the actually attained position/path of a robot end effector and the input position/path commanded by the robot controller
- Absolute Accuracy—A measure of difference between the actually attained position/path of a robot end effector and the input position/path (in absolute world coordinate frame) commanded by the robot controller (Figure 4.2)
- Relative Accuracy—A measure of difference between the actually attained position/path of a robot end effector and the input position/path (relative to an intermediate reference frame) commanded by the robot controller (Figure 4.3)

Accuracy can be defined in terms of "pose" accuracy, which relates to accuracy of static position/orientation, and "trajectory" accuracy, which relates to actual trajectory and commanded trajectory. The term "palletizing" accuracy is sometimes used to refer to accuracy of interpolated intermediate points between two taught points, which are assumed to be accurate. The term "teaching" accuracy is sometimes used to refer to a measure of difference between the actually repeated position and the taught position (taught by jogging the robot in teach mode or by lead-through teach method).

Accuracy is not the same as repeatability. While repeatability deals with

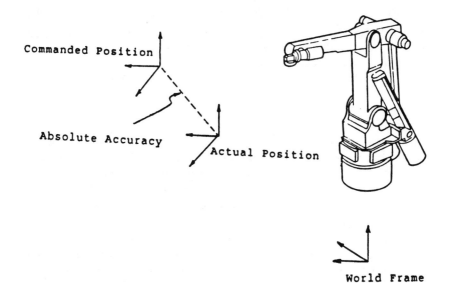

FIGURE 4.2.   Absolute Robot Accuracy

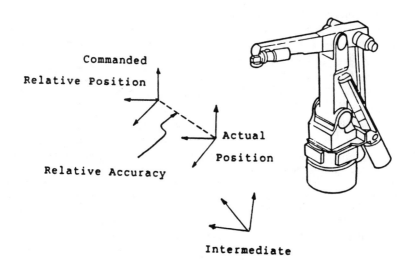

FIGURE 4.3.   Relative Robot Accuracy

positional discrepancies from cycle to cycle, accuracy deals with actual spatial positioning attained by the robot. Repeatability can be subclassified into omnidirectional repeatability and unidirectional repeatability [8, 9]. Unidirectional repeatability is measured with the robot moving from the same direction all the time. It is always smaller than the omnidirectional repeatability.

Repeatability is a function of many variables, such as the number of cycles, location in workspace, temperature, speed/acceleration of path, payload, etc. Like accuracy, there is no unified method of measuring repeatability. JIS (Japan Industrial Standards) uses a seven-cycle measurement. Large robot users, such as General Motors and Ford, have defined their own test procedures to measure repeatability. Other methods of tests and specifications, such as the use of measurement cubes [10] and spheres [11], have been proposed.

Since there are many factors influencing repeatability and accuracy, specifications of repeatability and accuracy are often creatively stated by manufacturers and perhaps unrealistically demanded by users. A common understanding of test conditions for repeatability and accuracy will be helpful in planning realistic applications. To have a common understanding of accuracy issues, factors affecting robot accuracy are discussed below.

3. FACTORS AFFECTING ROBOT ACCURACY

Robot accuracy can be influenced by a number of factors. They are classified into the following categories:

1. "Environmental" factors, such as temperature, humidity, and electrical noise,
2. "Parametric" factors, including kinematic parameters, such as robot link lengths, joint zero-reference angles, dynamic parameters such as structural compliance, drive-train compliance, friction parameters, varying inertia, and other nonlinearities, such as hysterisis, backlash, etc.,
3. "Measurement" factors, such as resolution and nonlinearity of encoders, resolvers, etc.,
4. "Computational" factors, such as robot-path computation errors from digital computer round-off, steady-state control error, tracking control error, etc.,
5. "Application" factors, such as installation errors, part presentation errors, errors in defining workpiece coordinate frame, etc.

Each category of inaccuracy may not be totally independent from one another. For instance, temperature change affects link length, friction coefficients, drifts in control electronics, and measurement sensitivity.

3.1. Environmental Factors

Temperature variations can cause link length changes in structural components and drive-train components. For instance, a change of ten degrees Fahrenheit (typical from day to night in plant environment) will cause a 1,000mm aluminum structure to change (by 1,000mm × 0.13E-4 × 10 = 0.13mm). The length of a 600mm steel ball screw can change by (600mm × 0.08E-4 × 10 = 0.048mm). If the distance of end-effector is eight times that of the moment arm of the ball screw, actual change of end-effector accuracy can be (0.048mm × 8 = 0.384mm). In many applications, this change will not impact productive use of robots. In other applications requiring precision, this is not tolerable. Temperature affects long-term repeatability and accuracy. It needs to be carefully considered in robotic applications.

Temperature and relative humidity affect characteristics of lubricants used in drive trains. Common lubricants used in robots are non-Newtonian. Their rheology behaves depending on how they have been used. The amount of lubricants on the drive elements varies depending on design and service conditions. It is normally difficult to have a good measurement of frictional characteristics in working industrial robots.

Temperature variations affect dimensions of rotating parts of rotary drives. These, in turn, cause changes in stiction, Coulomb friction and viscous friction. Clearance in drive-train components and hysterisis characteristics are affected. Such friction-related factors affect accuracy of robots and they cause difficulties in control.

Fiber-reinforced composite material are sometimes used in robot structures. By combining fibers of various coefficients of thermal expansion, an arm can be designed, so that there is little effective change of arm length within a range of temperatures. This is one possible solution in designing extremely accurate robots. However, resins used in fiber-reinforced composite material are also subject to changes in the chemical composition of the working environment of the robot. In extremely acidic or caustic environments, structural rigidity of the fiber-reinforced composite arm can suffer.

Typically, a robot controller contains a number of analog parts, such as resisters, capacitors, and operational amplifiers. There can be slight drifts in characteristics of these electronic components. However, they may contribute to several "counts" of offset values in each joint of the robot when ambient temperature varies or electrical noise is randomly injected. For a typical resolution of 0.001 mm per encoder count, a drift of ten counts can mean 0.010mm at the end effector. The cumulative drift from several joint axes can be several times greater than 0.010mm.

3.2. Parametric Factors

Robot parameters include kinematic parameters and dynamic parameters. Inaccurate representations of robot parameters contribute to robot inaccuracy. Generally, kinematic parameters affect "pose" accuracy and dynamic parameters affect "trajectory" accuracy.

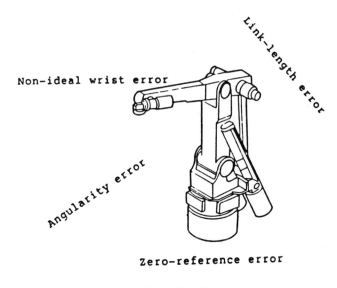

Non-ideal wrist error

Link-length error

Angularity error

Zero-reference error

Gear Train error

FIGURE 4.4.   Robot Kinematic Parameter Errors

3.2.1.   Kinematic Parameters.

The most noticeable kinematic parameters are robot link-length parameters (Figure 4.4). Link lengths are normally specified by design. They can vary due to environmental factors, and due to manufacturing and assembly errors. A link length can be different from its nominal value due to machining tolerance and tolerance stack-ups from a number of subassemblies. For each machined piece of a structure, machining tolerance is generally around $+/-0.050$mm. If a robot link were to be assembled from three pieces fastened together, overall length of the link, due to tolerance stack-ups, could vary by $+/-(0.050$mm $\times$ SQRT(3) $=$ 0.085 mm) RMS. For a three-jointed robot, end-effector position could then vary by $+/-(0.085$mm $\times$ 3 $=$ 0.225mm), which would be a relatively significant amount of error for many robot applications. Quality inspection and retrofit at the final stage of machining will help reduce such error. It would add cost to the robot components if accuracy were so desired.

Kinematic parameters include offset distances between each pair of adjacent joint axes. Offset distances can be different from nominally designed values by a similar amount as indicated in the last paragraph. In assemblage of robot manipulators, some link lengths or drive-shaft lengths are slightly modified to make all of the pieces fit together. With careful design and precision machining, many of the "fitting" tasks can be reduced in subassemblies, but not totally eliminated in the final assembly of the entire machine.

Angularity between each pair of adjacent axes also contributes to inaccuracy of robots. Angularity can come from assembly error of fitting subassemblies together or in the seating of bearings supporting the axes. An error of 0.1 degree in angularity could cause a link offset error of 1.745mm when the two adjacent axes were at 1,000mm apart.

Many industrial robots in use today adopt the design of the so-called "ideal" robot [12, 13]. "Ideal" robots have closed-form inverse solutions, such that desired Cartesian motions of robot end-effectors can be converted to joint angle commands directly. This allows fast computation in the robot controller, so that robots can follow desired programmed paths in real time. Typical "ideal" robots contain three adjacent axes of rotation that intersect at a common point or at infinity (in this case, the three adjacent axes are parallel). Realistically, errors caused by kinematic factors, due to manufacturing and environmental factors, often make "ideal" robots nonideal, thus, affecting accuracy.

The other kinematic parameters that contain inaccuracy are the zero-angle positions of robot joints. The zero-angle position of each joint can contain error due to imprecise seating of the encoder and resolver, or improper procedure determining the zero-angle position. As stated before, an error of 0.1 degree in defining a zero-reference angle can cause an inaccuracy of 1.745mm at 1,000mm away from the center of that particular axis. Whereas, all the kinematic parameters contribute to errors in accuracy, they, generally, do not contribute errors in repeatability.

Drive-train errors include nonuniformity of gears, chain, and belts in their range of motion. They cause inaccuracy in the workspace of robots. By careful measurements of gear-train motions, gear-train error can be mapped in robot software for accuracy improvement [14].

3.2.2. Dynamic Parameters.

Structurally, robots are designed to be more flexible than machine tools. Otherwise, robots can become heavy and inefficient. Being flexible, robots will deflect under load and under its own weight. Dynamic parameters, such as force, gravity, and inertia of robots are important in controlling accurate robot moves.

There are several components in robots that contribute to flexibility (i.e., link structures, bearing, and drive-train components). Links are usually designed with a cross-section, either box-shaped or I-shaped, to maximize their sectional module of elasticity. Bearings are usually designed not only to carry loads at desired speeds with given life, but also to have small angular deflection under moment loading. Drive-train components are designed with overall stiffness to be within a reasonable value. Each one of these three components (structure, bearing, and drive train) contribute as significant flexibility as the other two depending on the particular design of the robot mechanism. There is no fixed rule in determining flexibility values for industrial robots. They all contribute to robot inaccuracy when payload changes occur at the robot end-effector.

Inertial parameters of robots play a role in trajectory accuracy and velocity accuracy of robots. Since the power drives of robots, such as electric motors, do not provide instantaneously infinite power and they are torque limited, inertial parameters contribute to centripetal, tangential, and Coriolis forces that make robot controls challenging. Inertial parameters usually are not precisely measured in robot manipulators. Castings of structural parts could vary in thickness. Machining of castings could be offset if the starting reference dimension is offset. Heat from welding could cause slight distortion of parts. All such factors cause variations of inertial parameters from their originally designed values. Inertial parameters could vary by ten percent or more from robot to robot of the same model. Usually, measurements of inertial parameters are difficult and costly. Off-diagonal terms of the inertia matrix are especially difficult to obtain. Inaccurate inertia parameters can cause control errors. With inertia (J) and flexibility ($1/K$) in a dynamic system, we have resonance or natural vibration. Every machine has resonances, and ever robot shakes at, or near, its natural frequencies. Typical industrial robots have first resonance at about 20 hz, and a large majority of industrial robots have their first harmonic at less than 5 hz. Depending on the speed and direction of the drives, resonances may cause problems in accuracy and in the life of the components.

Friction parameters of robots are difficult to quantify accurately. In a typical nondirect-drive gear train, only 50 percent to 90 percent of the power is delivered from the output of the motor shaft to the, finally, driven joint. The rest is lost through friction as heat. Proper lubrication reduces friction, improves efficiency, and increases the life of the drive-train components. Viscous friction is one part of the many forms of friction. Commonly, in the robot drive system, stiction is normally evident, especially in large drives carrying a high payload. Hysterisis shows up when the drive is cycled back and forth, thus, contributing errors in omnidirectional repeatability and accuracy. Coulomb friction is a function of the normal force exerted between two contacting surfaces. All friction parameters dissipate energy and contribute to robot inaccuracy. They cause errors in dynamic identification trials.

3.3. Measurement Factors

Encoders and resolvers are the most common position feedback devices for robots. Other position feedback devices include inductosyn, LVDT, etc. They provide resolutions from several hundred counts to several hundred thousand counts per revolution. Encoders and resolvers can be mounted directly to the motor shaft or to the joint directly. Encoders mounted directly to the motor provide ease of control of the motor, since motor commutation is controlled directly. However flexibility, backlash, and other parametric factors affecting robot accuracy are not directly fed back to the robot controller. Thus, joint axes will require high resolution to detect small increments of joint positions. Compliance of the drive train between the motor and the rotated joint can affect robot performance and accuracy, especially at high payloads.

Sensors, such as strain-gage devices, can be mounted at appropriate places on robot links to measure strain changes, so that robot flexibility can be accurately controlled. End-effector inaccuracy could be reduced by using such methods. Load-sensing helps monitor machine tool conditons [15]. It detects loads, which can be used for flexibility compensations. However, all sensors are resolution-limited. If accuracy were to be improved, resolution of measurement devices would need to be better than the required accuracy.

External measurement instruments (not attached to the robot mechanism), such as Laser interferometer, theodolite, opto-camera system, and coordinate measurement machines, can detect robot end-effector accuracy away from the robot [16]. They provide reliable measurements of robot accuracy independent of robot parameters. Their own resolution and accuracy demands are high. Thus, they provide more accurate data than those from sensors on motors. Some of the measurement instruments provide static measurements only. Some sensors have a fast data sampling rate and have the capability to be integrated into robot controllers, so that accuracy information can be constantly fed back for real-time control. A method to perform on-line accurate robot position tracking using an external sensor integrated with a robot controller is the Laser tracking system developed by Lau. (Figure 4.5)

Table 4.1 shows some typical external measurement capabilities and limitations.

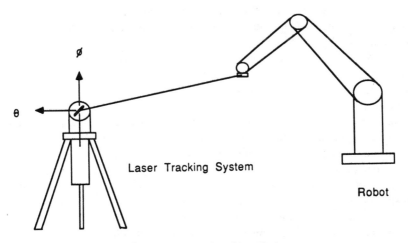

FIGURE 4.5.   Laser Tracking System

TABLE 4.1

	Resolution (with 2000mm range)	Limitations
Laser Tracker Interferometer	.1mm or better	Sensitive to light vel. in air, dust
Laser Interferometer	.02mm or better	Sensitive to light vel. in air, dust One-dimensional only
Theodolite	.1mm or better	human meas. error target size limit. low manual process
Opto-camera system	1mm or better	camera resolution
Coordinate Measurement	.02mm or better	immobile confined workspace
Collimated Sensor System	.4mm or better	small work volume

3.4. Computation Factors

The desired path of the robot end-effector is calculated in the robot controller. Precise commands to individual joints are then issued. This path computation takes time. In today's industrial robots, it can range from several milliseconds to several hundred milliseconds. Computations of paths contain several types of errors. The computer controller uses a finite number of bits to represent numbers. For most computations, 16 bits will be enough to retain precision in robot path computations. For computations that contain the accumulation of small increments of motions, there can be inaccuracy due to round-off errors.

For instance (Craig, Chapter 3 [17]), a motor shaft attached to a ball screw may have a resolution of 1,000 counts per revolution. This ball-screw is used to rotate an arm hinged at one end. If the pitch of the ball screw is 10mm per revolution, finite resolution of the ball-screw position is then 0.01mm per count. Since the length of this ball screw is used to control the angular motion of the arm, actual resolution of the angle spanned by the ball screw is about $(0.01mm/La)$ radian. If La (La: length of moment arm) is 250mm, actual resolution of the angle is about 0.00004 radian per count. For a total range of travel of 1.2 radian, a total of 30,000 counts is required, enough for a 16-bit representation of angles.

However, the actual angle computed from the ball-screw travel comes from the cosine formula. From computations of square roots and arccosine, it is likely to lose one count of theta, due to round-off. After several cycles of the ball screw, actual position of the robot arm can be off by several counts. This example is meant to illustrate the general round-off or truncation error. Increasing the number of bits will help reduce such round-off error.

Another factor contributing to the inaccuracy of robots is the case of the "nonideal" arm. Typical computations of inverse solutions use direct formulations, which assume ideal robot kinematics. This error is in the range of the manufacturing tolerance, which can be in the order of tenth of a millimeter. In order to reduce error, due to nonideal robot arm, robot kinematic computation needs to be done using nonideal formulations. Nonideal formulations require much more computations than today's robot controllers can handle. Algorithms using numerical iterative approaches to handle nonideal robot configurations can be developed, but real-time general solutions are needed to overcome the error due to "nonideal" robots.

Robot positions at, or near, robot singularities present challenges in computations. There are many ways to handle singular positions in robot manipulators. Very often, accuracy is slightly compromised near such singular positions in order to achieve stable motions.

3.5. Application Factors

Besides the above-mentioned factors affecting robot accuracy, actual applications of robots also contribute to inaccuracy of the system. The main reason for demanding accuracy in the first place is the need to perform the job or application. Besides accuracy of robots, factors related to the application need to be looked into.

A common error happens in robot installation. Some robots are installed on rather soft surfaces. Some are installed by grouting on the cement floor. Without adequate dimensional control of robot installations, the world-coordinate frame, by which all other coordinates are based, will contain error.

The next factor relates to the dimensional integrity of the workpiece. Due to tolerance stack-ups, and other manufacturing errors on the workpiece, the demand to perform applications is often unduly placed on the robot. It may just be a simple correction to the workpiece design and manufacturing process to allow the entire robot system to perform accurately.

End-of-Arm tooling inaccuracy is another factor contributing to the overall robot inaccuracy. Like manufacturing errors in robot parameters, tooling design/manufacturing/installation also contain errors. Sometimes it may be less costly and more effective to improve the accuracy of end-of-arm tooling than the robot.

4.   METHODS OF IMPROVING ROBOT ACCURACY

After looking over the factors affecting robot accuracy, it may seem alarming that robots can be quite inaccurate. It is generally true that inaccuracy exists, but there have been many methods developed by users, manufacturers, and researchers to make improvements for the intended applications. Quality inspection, with particular attention on accuracy, will reduce errors in manufacturing and assembly. Providing an environment for consistent robot accuracy will definitely help.

Robot users and manufacturers will constantly make improvements in robot accuracy. Some of the methods for improving robot accuracy are discussed in the following three categories:

1. Calibration methods, including many of the existing methods of calibration, mastering, and reference matching techniques. They generally improve robot accuracy at only one particular point in the robot workspace. We may term this to be a "first-order" improvement in robot accuracy.
2. Open-loop methods, including many of the robot parameter identification methods in use or in development, deal with kinematic parameter identification. Drive-train errors and compliance errors have also been investigated with good results. Open-loop methods depend on off-line measurements to identify the necessary parameter corrections. After identification of errors, the corrected parameters are used by robot controllers in open-loop computation of robot paths. These methods provide another order of improvement over the calibration method, as accuracy for the robot workspace can be enhanced. Dynamic identification and control require computations that are not yet economically attainable in most industrial robots, even though they can improve trajectory accuracy.
3. Closed-loop methods, including methods that incorporate external sensors that detect end-effector positions, are integrated on-line with robot controllers. The positions of the end-effector are fed back to the robot controller on-line and an additional control loop modifies the robot paths. The Laser tracking method is an example of the closed-loop method. Besides the above-mentioned compensation methods, some special end-of-arm tooling developments have worked successfully in robot assembly applications. Even though the remote-center compliance devices and general-force sensors do not directly improve robot accuracy, they work reasonably well in dealing with the inaccuracy of robots and parts [18].

4.1. Calibration Methods

4.1.1. Robot Calibration.

Robot calibration is a procedure to establish zero-reference positions of robot joints. It determines the relationship between encoder values and actual robot-joint angles. Calibration must be done in order to command accurate robot moves in Cartesian frames. There are several methods to accomplish robot calibration. As a common rule, a predetermined robot pose is selected. This pose can be established using external sensors or, more frequently, by physical markings on the robot structure. Some robots use microswitches tripped by cams/dogs to determine the particular pose. After the pose is attained, a record command is issued to establish a common reference between encoder values and actual joint values.

Some robots utilize absolute encoders/resolvers as feedback of joint positions. Absolute encoders eliminate the need for users to go through the robot calibration process. It then becomes the manufacturer's responsibility to ensure the initial determination of zero-reference positions.

Using scribe marks or dogs may not be consistent from robot to robot. Therefore, a fixture is sometimes used to establish a common position of the dogs/cams. Sometimes the fixture is used directly as part of the calibration procedure. This fixture is commonly referred to as a calibration fixture or a mastering fixture. Depending on the design of robot mechanisms, the precision mastering fixture can also determine the adjustments needed on robot lengths in addition to providing reference for the joint zero angles. The calibration/mastering procedure helps establish accuracy at one particular position in the robot workspace. It is one of the first-order corrections to improve robot accuracy.

Robot calibration can normally be performed within $+/-0.025$mm accuracy at that one particular point in the robot workspace.

4.1.2.   Workspace Calibration.

As indicated in Figure 4.1, there are several reference frames used in describing a robot system. In order to perform the intended task accurately, the workpiece or station frame must be accurately established. The workpiece frame can be measured using external sensors with respect to a world frame. Since robots perform tasks directly in the workpiece frame, direct transformation from robot frame to workpiece frame can be sufficient. One common method to establish relative accuracy between a workpiece and robot is to command a robot to move to three calibrated positions on the workpiece. A 4×3 transformation matrix is then calculated to transform from robot coordinate frame to the needed workpiece frame. This transformation calculation is direct and fast, even though robot parameter errors still remain uncorrected.

4.1.3.   Sensor Frame Calibration.

In cases of sensor-guided robots, a sensor-reference frame needs to be established. Sensors, such as vision cameras, can be mounted on the robot or at a fixed position away from the robot. As shown in Figure 4.1, a sensor frame needs to be established in order to perform accurate robot tasks. There are many methods in which sensor frames can be established with respect to the robot or with respect to the workpiece. This is one of the challenging areas in sensor-guided robot applications.

4.2.   Open-Loop Methods

Based on some predetermined information about the robot and factors contributing to inaccuracy, the open-loop methods make compensated commands to the

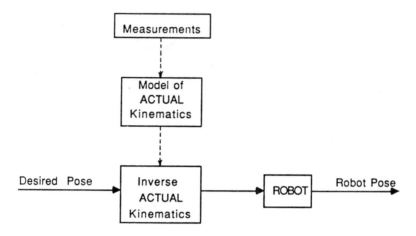

FIGURE 4.6.   Open-Loop Correction Method

robot, so that robot end-effector accuracy can be improved. Such accuracy improvement is intended to work in most of the robot workspace, rather than one particular calibrated point.

Robot inaccuracy (absolute) has been reported to range from five mm to ten mm in the robot workspace for typical robots [19, 14]. Unpublished robot inaccuracy, measured by users and manufacturers, generally agree with the reported figures. Several methods have been proposed to make improvements. They are classified here as open-loop methods.

Three categories of tasks are involved in the open-loop methods (i.e., measurement task, modeling task, and compensation task). As indicated in Figure 4.6, measurements and modeling are done off-line to identify the needed compensations. Actual compensations are done on-line while commanding robot moves.

4.2.1.   Measurement Tasks.

Robot accuracy measurement is the first step in understanding what errors need to be corrected. Two sets of measurements are needed. One set contains robot end-effector positions measured by an external sensor (e.g., a theodolite system). The other set contains the corresponding robot positions as commanded by the robot controller. With proper transformation between the reference frames, actual errors in robot accuracy can be measured.

An automated method of measuring robot accuracy has not been widely applied. The reported measurements on robot accuracy required a great deal of effort in data (taking from several man-days to several man-months.) Even though part of the time was sent in checking the integrity of data, it was generally recognized as a rather tedious task. Several methods have been proposed by developers to facilitate robot accuracy measurements.

4.2.2. Modeling Tasks.

After obtaining robot accuracy data, analysis is needed to determine the type of corrective measures to be made. Several algorithms have been proposed by researchers, from purely kinematic parameter corrections [20, 21, 22, 23, 24] to full models including gear-train and backlash corrections [14, 19].

Generally, error modeling assumes small parameter errors, so that correction parameters become linear in the equations. Parameters to be corrected are then calculated by using least-square fit or by minimizing a global-error functional. With corrections of robot parameters, robot inaccuracy can be reduced from five to ten times to about one mm in the workspace [19]. It becomes increasingly difficult to achieve better than one mm (for large robots) since many more parameters, such as load-deflection models, are needed. With better understanding of robot mechanisms and more parameters identified, accuracy improvement can be made much better theoretically. Small robots can be made more accurate from using a good set of identified parameters, but it is extremely difficult to achieve anything better than 0.5mm accurately in the entire workspace.

4.2.3. Compensation Tasks.

There is a limit to how many parameter compensations the robot controller can actually handle. From knowledge of mechanical design of the robot, critical parameters to be corrected can be isolated out of a set of general parameters, thus, reducing the identification task and compensation task.

In open-loop methods, the corrected parameters are stored in the robot controller. Robot motions are then commanded based on the corrected parameters without additional feedback of robot accuracy. Compensation tasks can be easily done with ideal robot configurations. If there were parameters that would render the robot nonideal, computations would become increasingly complex. Parameters, such as offsets, may require computations of robot Jacobians as part of path planning. In most industrial robots today, it is still not economically practical to compute Jacobians on-line. In the near future when computation power of robot controller improves, this may not be a problem. Dynamic parameter identification cannot be fully done by measuring robot positions statically. Dynamic identification and control is still very much in a research stage.

4.3. Closed-Loop Methods

Using parameter identification and then compensating robot parameters do not provide exact positioning information of the robot. As discussed in the previous section, robot parameters are identified using a "best-fit" approach to the data taken at various locations. There is still error left at each measured location and likely more so at locations elsewhere in the workspace. The next improvement to achieve a positioning accuracy in overall robot workspace is to use a feedback method where the robot positions are constantly monitored using a external

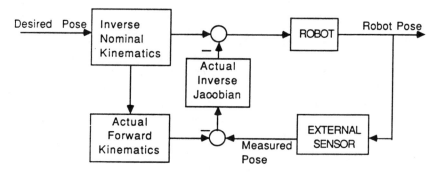

FIGURE 4.7. Closed-Loop Correction Method

sensor [25]. As shown in Figure 4.7, the robot position is controller closed-loop with the externally fedback positions. The Laser tracking system measures the robot end-effector position with accuracy of 0.012mm in a two meter workspace [16]. This information can be fedback to the robot controller for closed-loop control of robot accuracy. This is about the best accuracy one can expect today when integrating such a Laser tracker with a robot controller. A sketch utilizing the closed-loop concept is shown in Figure 4.8.

5. CONCLUSIONS

In this chapter, factors affecting robot accuracy are identified. Several methods to improve robot accuracy are discussed. It is important for users of robots to

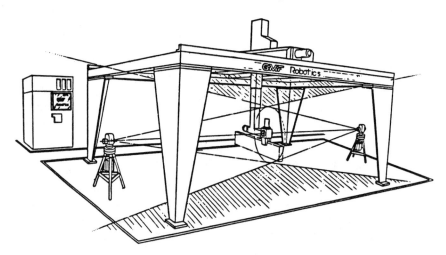

FIGURE 4.8. A Closed-Loop Laser Tracking Robot System

understand the factors affecting accuracy and where present technology stands in making accuracy improvements. There are intrinsic limitations in robot manipulators, and in robotic applications. It will require joint efforts from users, manufacturers, and researchers to achieve the most productive use of accurate robot systems.

REFERENCES

1. D.L. Pieper, and B. Roth, "The Kinematics of Manipulators Under Computer Control," *Proc. International Congress on the Theory of Machines and Mechanisms,* Vol. 2, 1969, pp. 159–168.

2. D.E. Whitney, W.J. Book, and P.M. Lynch, "Design and Control Considerations for Industrial and Space Manipulators," *JACC,* 1974, pp. 591–598.

3. A.K. Bejczy, "Performance Evaluation of Computer-Aided Manipulator Control," *Proc. of the 1976 IEEE Int. Conference on Cybernetics and Society,* Washington, DC, November 1–3, 1976.

4. B. Roth, J. Rastegar, V. Scheinman, "On the Theory and Practice of Robots and Manipulators," *First CISMIFToMM Symposium,* Vol. 1 September 5–8, 1973, pp. 93–113.

5. C.P. Day, L.L. Whitcomb, D.J. Daniel, "Grayscale Vision System for Robotic Applications," *Proc. Robot 9 Conference,* Detroit, MI, June 4–7, 1985.

6. M.P. Jacobs, "Offline Robot Programming: A Current Approach," *Proc. Robots 8,* Detroit, MI, June 4–7, 1985, pp. 4.1–4.7.

7. K. Lau, N. Dagalakis, D. Myers, "An Overview of Techniques for Robot Performance Measurements," Center for Manufacturing Engineering, National Bureau of Standards, Gaithersburg, MD, 1986.

8. J.C. Colson, and N.D. Perreira, "The Need for Performance Measures for Robotic Systems and Some Qualitative Definitions," *Workshop on Robot Standards, co-sponsored by ASTM, ANSI, IEEE, and IEQ,* Detroit, MI, June 6–7, 1985.

9. J.C. Colson, and N.D. Perreira, "Quasi-static Performance of Robots," *Robotics and Computer Integrated Manufacturing,* Vol. 2, No. 3–4, Pergamen, NY, 1985, pp. 261–270a.

10. P. Ranky, "Test Method and Software for Robot Qualification," *The Industrial Robot,* June 1984.

11. B.W. Mooring, and C.R. Tang, "An Improved Method for Identifying the Kinematic Parameters in a Six Axis Robot," *Proc. of the 1984 Int. Comp. in Eng. Conf.,* Las Vegas, NV, August 1984, pp. 79–84.

12. B.K.P. Horn, "Identification and Compensation of Robot Kinematic Errors," 1983.

13. R. Paul, *Robot Manipulators: Mathematics, Programming, and Control,* M.I.T., Cambridge, MA, 1981.

14. D.E. Whitney, C.A. Lozinski, J.M. Rourke, "Industrial Robot: Calibration Method and Results," *Proc. of the 1984 Int. Comp in Eng. Conf.,* Las Vegas, NV, August 1984, pp. 92–100.

15. S.K. Birla, "Sensors for Adaptive Control and Machine Diagnostics," *Machine Tool Task Force Study—Machine Tool Controls,* Vol. 4, SME, Dearborn, MI, October 1980, pp. 7.12.1–7.12.70.

16. K. Lau, R. Hocken, L. Haynes, "Robot Performance Measurements Using Automatic Laser Techniques," *NBS-Navy NAV/CIM Workshop on Robot Standards,* Detroit, MI, June 6–7, 1985.

17. J. Craig, *Introduction to Robotics: Mechanics and Control,* Addison-Wesley, MA, 1986.

18. S. Drake, "Using Compliance Instead of Sensory Feedback for High Speed Robot Assembly," *6th North American Metalworking Research Conference,* April 16–19, 1978.

19. J. Chen, and L.M. Chao, "Position Error Analysis for Robot Manipulators with All Rotary Joints," *Third IEEE International Conf. on Robotics,* March 1986, San Francisco, CA, pp. 1011–1015.

20. C.H. Wu, "The Kinematic Error Model for the Design of Robot Manipulator," *Proc. of the 1983 American Control Conf.,* September 1983, San Francisco, CA, pp. 497–502.

21. H.W. Stone, A.C. Sanderson, C.P. Neuman, "Arm Signature Identification," *Third IEEE Int. Conf. On Robotics,* March 1986, pp. 41–48.

22. T.W. Hsu, and L.J. Everett, "Identification of the Kinematic Parameters of a Robot Manipulator for Positional Accuracy and Improvement," *ASME Proc. of the 1984 Int. Computers in Eng. Conf.,* August 1984, pp. 263–267.

23. D. Payannet, M.J. Aldon, A. Liegeois, "Identification and Compensation of Mechanical Errors for Industrial Robots," *Proc. 15th Int. Symp. on Industrial Robots,* Tokyo, Japan, September 1985, pp. 857–864.

24. B.W. Mooring, and T.J. Pack, "Determination and Specification of Robot Repeatability," *Third IEEE International Conf. on Robotics,* March 1986, San Francisco, CA, pp. 1017–1023.

25. M. Tucker, and N.D. Perreira, "Motion Planning and Control for Improved Robot Performance," *Int. Industrial Controls Conf.,* Long Beach, CA, September 16, 1986, pp. 1–17.

5

Intensity Blending of Computer Image Generation-Based Displays

E.A. Reidelberger
McDonnell Douglas Corp
St. Louis, MO

Daniel C. St. Clair
University of Missouri-Rolla
Graduate Engineering Center
St. Louis, MO

1. INTRODUCTION

Air combat simulators provide vital training for the aircrew of fighter aircraft. It is critically important to provide the most realistic simulation possible for training crews in air combat maneuvering. Air combat simulators are used to complement "real world" training [1]. Critical skills can be developed in simulators without risk of life or equipment, and for a fraction of the cost of actual flight time. Training in this manner can make actual flight time more valuable and more productive [2]. These goals require that the simulator provide the best possible training for the aircrew.

A state-of-the-art air combat simulator consists of a cockpit enclosed by a projection surface on which a visual scene is shown. As the pilot operates the aircraft controls, the visual scene responds. To provide a highly realistic and dynamic background, it is imperative that proper cues be provided which will elicit the correct response from the trainee [3]. The combined effects of high speed and low altitude make it a technical challenge to change the visual scene and apparent motion of the simulator realistically in response to the pilot's actions. A high-quality computer-image generation (CIG) system produces superb quality real-world images which provide the flexibility to enhance training with realistic sky/earth background scenes. The result is an accurate, meaningful visual training environment that can meet the complex military needs for simulated visual scenes in simulators.

2. COMPUTER-GENERATED IMAGERY

Computer image generation systems use a digital object/terrain database to compute geometric images of real-world scenes. The system consists of equipment to
generate, in a visual scene relative to a current viewpoint, a variety of terrain and
person-made cultural features [4]. Imagery is generated in a video format which
is usually a 1023 line raster that can be projected on a screen by special purpose
light valve projectors. The images are viewed from a design eye with six degrees
of freedom. These images are updated rapidly enough to accurately simulate
translational and rotational motion cues. Potentially visible scenery is extracted
by processing the database, an organized hierarchy of models, objects, and faces
against the dynamic data defining the position and attitude of the viewpoint. The
environment, containing moving models and terrain environment models, is
scanned to select the models that are viewable within the display channel boundaries. The models are projected into the viewplane and transformed to 2-D
projections as observed from the viewpoint. The 2-D information is filtered to
eliminate that portion which is imperceptible in size or is hidden from the
viewpoint.

2.1. Wide Angle Visual Displays

The goal of realism in visual displays has led to the use of wide angle displays
which are projected onto spherical screens, or domes. Domes are used to overcome the 180 degree limitation of flat screens by allowing several projectors to
project imagery onto different areas [5]. A dome provides full 360 degree visual
coverage. It offers wrap-around background visual imagery of earth and sky
scenes which provide distinct, unambiguous altitude, attitude, and velocity cues.
This is particularly critical when dealing with flight simulators for training pilots
to carry out low-level flying, ground attack, air-to-air combat, and other missions
requiring critical maneuvering and visual tracking [6]. Light valve projectors
fitted with specially developed optics packages can place an image on an area
greater than a 180 degree field of view. Hence, only two light valve projectors
are required for full 360 degree coverage. This approach provides a highly
realistic display with an unrestricted field of view. A typical domed air combat
simulator is shown in Figure 5.1. Note that the aircrew has unrestricted vision.
Anything visible to the aircrew in the real aircraft with the canopy closed and
head motion limited to that experienced by the aircrew wearing helmets and
unlocked shoulder harnesses is visible to the aircrew in the dome.

   The cockpit itself restricts some vision due to its structure. For example, the
aircrew is unable to see below the aircraft because of the airplane structure.
Projectors "hidden" in these areas can display images on the dome surface, yet
they would be unseen by the aircrew. However, there is no location inside the

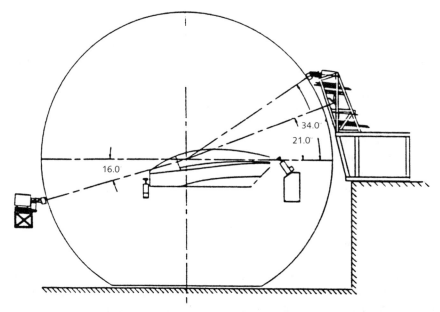

FIGURE 5.1.   Typical Domed Aircraft Simulator [7].

dome to mount a projector which will be out of the aircrew's field of view and still provide a full, unobstructed hemispherical display. Therefore, the light valve projectors must be located outside the dome. Each projects its image through a small hole in the dome. The positioning of the holes is done carefully to insure that they are invisible to the aircrew. Combining the hemispherical displays from the two light valves produces a full 360 degree display in the dome. The realism of the combined display depends on the hemispherical display continuity at points where the images overlap.

3.   HEMISPHERICAL DISPLAY CONTINUITY

The two hemispherical projected images must be joined together in a seam that is indiscernible. Any discontinuity between the separately displayed images results in an undesirable artifact. Two approaches for solving this problem are: (a) butting displays together, or (b) blending displays together.

3.1.   Butting Displays Together

Butting two hemispherical displays together is the simpler, though potentially more inaccurate, method of combining the two displays into one continuous display. If both hemispherical displays are adjusted to exactly the same color, it is theoretically possible to position the two light valve projectors so they are perfectly aligned. However, light valves change or "drift," with time, both in color and position. Light valve color output can be adjusted manually; however, due to drifting, the adjustment is not guaranteed to hold for any significant duration of time. Furthermore, the drifting experienced by light valves manifests itself differently for each light valve projector. The ambient conditions surrounding each projector can also add significantly to drifting.

Any movement of the projector or the video output from the projector, including the positional drift mentioned above, results in geometrical inaccuracies which produce discontinuous alignment of the two displays. These limitations prohibit this method from serving as a viable solution for providing hemispherical display continuity.

3.2. Blending Displays Together

The second method, intensity blending of the two hemispherical displays, suffers from none of the limitations encountered in display butting. Intensity blending can be handled optically or through software-controlled intensity modifications. A significant amount of special hardware must be designed and developed to blend two images together optically [8].

Using a software-controlled intensity blending algorithm requires modulation of video intensity signals along a blending band such that the result is a modified video signal wherever blending is to occur. Figure 5.2 shows the dome with the channel boundary surrounded by a three degree blending region. With two adjacent channels matched in luminance and color, the system will render the boundary between the two hemispherical channels nearly indiscernible to the aircrew. There will be no noticeable variation in luminance, color, or contrast across the boundary.

The blending of the two hemispherical channels is accomplished in the region formed by overlapping the hemispherical images. Hardware postprocessing is used to accomplish this blending. Figure 5.3 shows a diagram of this process. The system consists an IBM-PC/AT fitted with a specially developed graphics controller board. It also includes special purpose video multiplier boards mounted inside the IBM-PC/AT. These video multiplier boards, in conjunction with the graphics boards, perform the intensity blending. The blending system provides smooth matching of discontinuities at the boundary between adjacent fields by adjusting the intensity signals of each channel within the blending region. This adjustment is performed by a specially designed software algorithm which controls the video signal gains at the edges of the projection hemisphere.

The algorithm actually modifies the elements of a lookup table. The values in this table specify the intensity of each pixel in every raster line. The intensity of

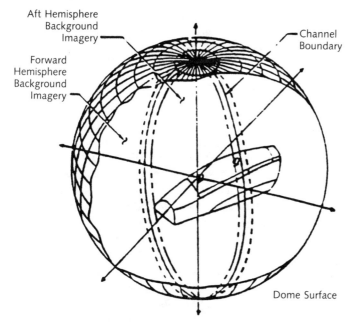

FIGURE 5.2. Dome Blending Region [7].

all (x,y) elements within the blending region will be modified according to the algorithm to be discussed in the following sections. The output of the table lookup is then used to control the intensity gains of the color video signals.

The actual data written to video memory consist of indices into a color pallet. Each entry in the color pallet contains three video-level data parameters, one each for red, green and blue. If all three color parameters are the same, a shade of grey from black to white is the result. The blending algorithm does not deal with colors separately, but with intensity. It is a prerequisite that the output of the

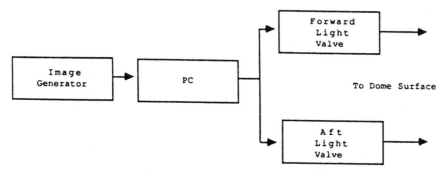

FIGURE 5.3. Blending Process Flow Chart.

projectors be matched in color by a manual adjustment of the projectors. By adjusting brightness the blending band covered by both displays will not appear brighter than the surrounding display.

4.   BLENDING REGION DEFINITION

To provide a region of overlap in which blending occurs, each of the two hemispherical displays must cover an area greater than 180 degrees. Three degrees was chosen as the width of the region. This width provides a sufficient region for blending yet a narrow enough region to minimize the additional images that must be generated during real-time operation. Using a three degree blending region requires that each hemispherical display cover 183 degrees. Figure 5.4 shows an expanded hemispherical display. The area covered within the inner solid line represents the 180 degree coverage. The outer solid line

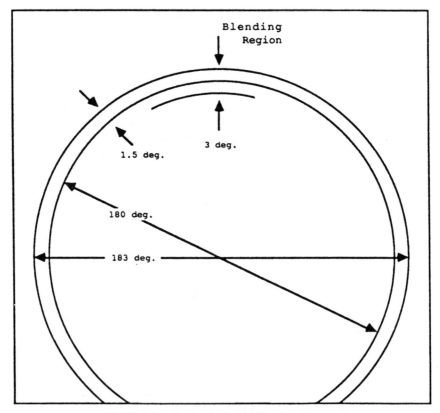

FIGURE 5.4.   Single Projector Display Coverage.

represents the extended 183 degree coverage which allows a one and a half degree overlap in all areas. When both displays are combined, a three degree blending region results.

By placing indicators of some sort along the equator separating the hemispherical displays, the ideal location of the 180 degree field of view can be determined. These indicators are placed at 10 degree intervals. Indicators are also placed on alternating sides of the centerline at one and a half degrees from the centerline, forming the three degree blending region as shown in Figure 5.5. The exact locations of the indicators are determined when they are installed in the dome. However, due to light valve drift, the perceived location of each indicator may not stay constant. If it were possible to assure that no effect from light valve drift would occur, blending would have to be done only once as the perceived location of the blending region would stay constant as long as the location of the image on the dome stayed constant.

During the adjustment phase, the CIG projects a pair of cross hairs on the dome. The intersection of these cross hairs is positioned over one of the indicators defining the blending region, and the coordinates are noted. The perceived azimuth and elevation of that indicator can be found and stored for use in later computations. This procedure is repeated for each indicator.

The blending region is subdivided into blending segments, each defined by two sets of indicators, one with center and an outer indicator and the other with center and an inner indicator as seen in Figure 5.5. After the perceived location of all indicators is known, the inner and outer boundary points associated with each centerline point, but not represented by an indicator, can be found. This is done by taking the vector from the existing boundary point, either inner or outer, to the associated centerline point, scaling the result by two and adding it to the existing boundary point to generate the imaginary boundary point. The result is an exact definition of the blending band including inner and outer boundary points and blending segments. Figure 5.6 shows these boundary point locations.

By forming the segments defined by each associated inner and outer boundary point, the boundary segments appear as shown in Figure 5.7. These segments allow for a more accurate definition of the blending elements. This definition depends on each segment's relationship to standard space.

5.   STANDARD SPACE

The definition of the blending region and each blending segment is used to determine how video memory must be modified to provide blending. Video memory has a one-to-one correspondence to raster space. Some method must be used to assure evaluation of each element that must be modified. Since realism in background displays requires a minimum resolution of 1024 × 1024 raster elements, individual element evaluation without regard to its relative location to

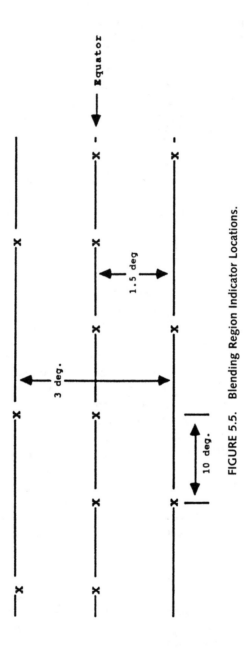

FIGURE 5.5.   Blending Region Indicator Locations.

116

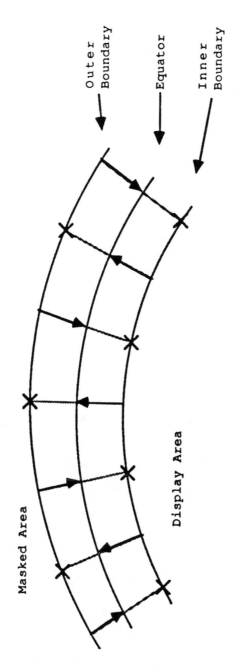

FIGURE 5.6. Boundary Point Locations.

117

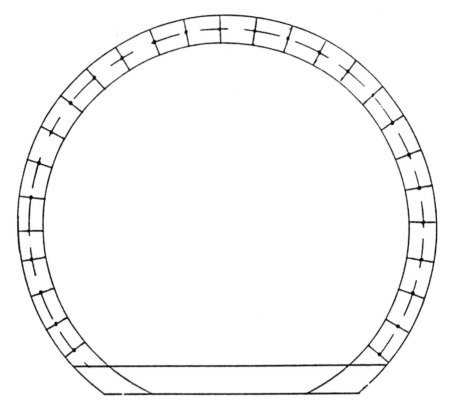

FIGURE 5.7.   Blending Region Segments.

other elements would be too time consuming. To simplify this procedure, a method was sought that could be used to provide the calculations necessary with the least amount of repetition. The procedure would use information about an element's location to reduce the intensity computation.

A standard space was defined as the common orientation into which each quadrant can be mapped. One unique transformation matrix for each quadrant maps that quadrant into standard space. The standard space concept is a very powerful component of the blending algorithm in that it helps expedite raster line evaluation.

By defining standard space as shown by the representative quadrant in Figure 5.8, the number of unique cases is greatly reduced. Further, the relationship of each raster line to the outer and inner boundaries can be defined. Any raster line in standard space may intersect the outer boundary at most once. When the outer boundary is intersected, the raster line may intersect the inner boundary at most once, as shown by raster lines A and B. The case in which the outer boundary is not intersected is trivial, as shown by raster line C.

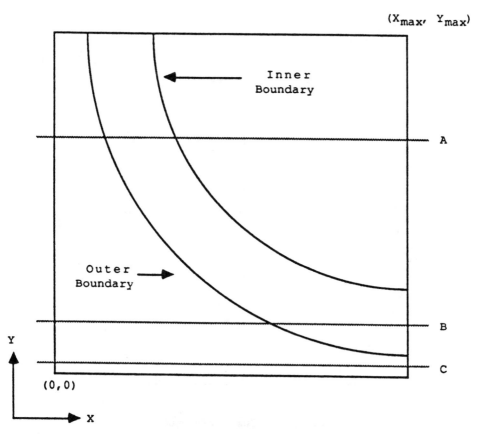

FIGURE 5.8.   Standard Space Representative Quadrant.

By dividing raster space into four quadrants as shown in Figure 5.9, the blending algorithm can be drastically simplified. Each element of raster space, (X_R, Y_R), can be mapped into standard space by the equation

$$[X \ Y] = C_i \begin{bmatrix} X_R \\ Y_R \end{bmatrix}$$

which uses a unique transformation, C_i, for each quadrant as shown below. The subscript i denotes the quadrant being referenced.

$$C_1 = \begin{bmatrix} 1 & 0 \\ 0 & 1 \end{bmatrix} \qquad\qquad C_2 = \begin{bmatrix} -1 & 0 \\ 0 & 1 \end{bmatrix}$$

$$C_3 = \begin{bmatrix} 1 & 0 \\ 0 & -1 \end{bmatrix} \qquad\qquad C_4 = \begin{bmatrix} -1 & 0 \\ 0 & -1 \end{bmatrix}$$

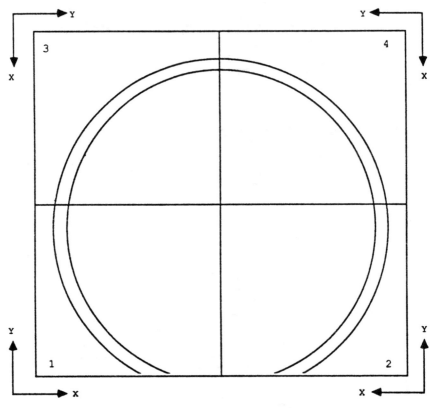

FIGURE 5.9.   Quadrant Definition.

In standard space, the blending region is positioned in the quadrant such that a raster line will intersect the outer boundary before intersecting the inner boundary, when evaluating it from left to right (see Figure 5.8). The area to the left of and/or below the outer boundary is the masked area, while the area to the right of and/or above the inner boundary is the display area. The blending region separates the masked area from the display area. Figure 5.10 represents each quadrant and its resultant orientation. The blending and raster fill processes always takes place after the quadrant has been mapped into standard space.

6.   RASTER LINE EVALUATION WITHIN THE BLENDING REGION

The concept of standard space facilitates the development of a uniform procedure for evaluating each raster line to determine its relationship with the blending

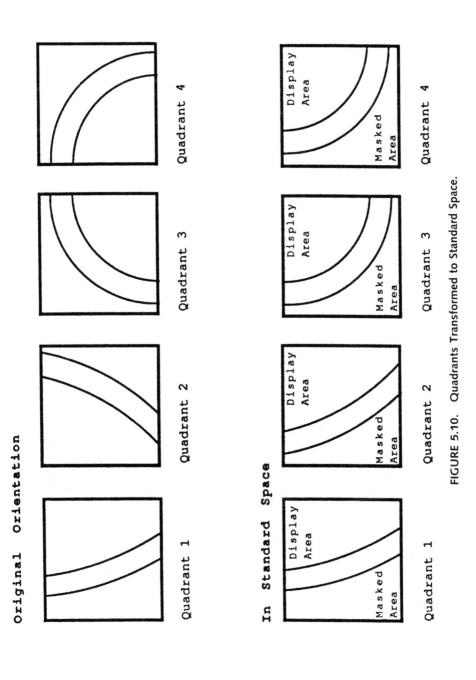

FIGURE 5.10. Quadrants Transformed to Standard Space.

121

region. Transforming each quadrant to standard space before performing raster line evaluation permits the same algorithm to be used on all raster lines. After evaluation, lines are transformed back to their original quadrant. Raster lines are evaluated from bottom to top, from left to right, beginning with the first raster line.

6.1. Evaluation of the First Raster Line

It is apparent that one of two situations occur for the first raster line evaluated:

1.   it intersects the outer blending boundary, or
2.   it does not intersect the outer blending boundary.

These two situations are shown in Figure 5.8.

To determine which situation applies, the intersection of the outer boundary with the vertical line $x = X_{max}$ must be determined. If the intersection falls below the quadrant, the first raster line will intersect the outer boundary. If not, the y-component of the point of intersection indicates the first raster line which intersects the blending region.

To calculate the point of intersection, observe that the outer and inner boundaries closely approximate circles. If the equation of the circle defining the outer boundary can be determined, computing the intersection of the raster line and the circle is trivial.

One way to calculate the intersection point is to use all outer boundary points in determining the equation of a single circle which would be used as an approximation to the outer boundary. Depending on the location of the outer boundary points, both perceived and imaginary, this approach would likely introduce a relatively large error into the calculations.

To improve accuracy, the equation of the circle approximating the outer boundary will be calculated using the three closest outer boundary points. While this requires multiple circle approximations, it does improve accuracy. The three points to be used to define the circle are the three outer boundary points closest to the Y value of the raster line. These points are labeled P_1, P_2, and P_3, as shown in Figure 5.11.

The circle's center (X_C, Y_C) is the point of intersection of the perpendicular bisectors of the line segments P_1P_2, and P_2P_3. The equation of the circle containing these outer boundary points is

$$(X - X_C)^2 + (Y - Y_C)^2 = r^2 \tag{1}$$

where

$$r = [(X_O - X_C)^2 + (Y_O - Y_C)^2]^{1/2}.$$

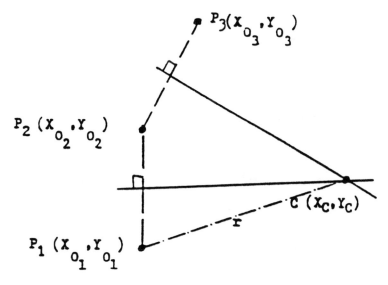

FIGURE 5.11.   Circle Definition From Three Boundary Points.

Point (X_O, Y_O) denotes any of the three known outer boundary points P_1, P_2, or P_3.

Having determined the equation of the circle, it is possible to determine the point (X_{max}, Y) where the circle intersects the vertical line $x = X_{max}$. Since the blending algorithm requires integer values, it may be necessary to truncate the Y value. Let the resulting value be denoted by Y_{min}. The last raster line not intersecting the outer boundary is $y = Y_{min}$. If $Y_{min} < 0$, all raster lines in the quadrant intersect the outer boundary. Otherwise, each pixel of raster lines with a value of Y_{min} or less will have an associated value of zero, indicating this area is masked out of the display. Because of the definition of standard space, every remaining raster line $y = Y_j$ where $Y_{min} < Y_j < Y_{max}$ and where Y_{max} is the value of the uppermost raster line, will intersect the outer boundary. In many cases it will also intersect the inner boundary.

6.2.   Evaluation of Succeeding Raster Lines

Each succeeding raster line after the first must be evaluated to determine into which one of three cases if falls (see Figure 5.12). Figure 5.12A shows blending segments formed by connecting the associated inner and outer boundary points. Figure 5.12B shows the three cases which can result as a raster line crosses the blending band. Case 1 occurs when only one segment is intersected. When more than one segment is intersected, either case 2 or case 3 exists. The slope of the

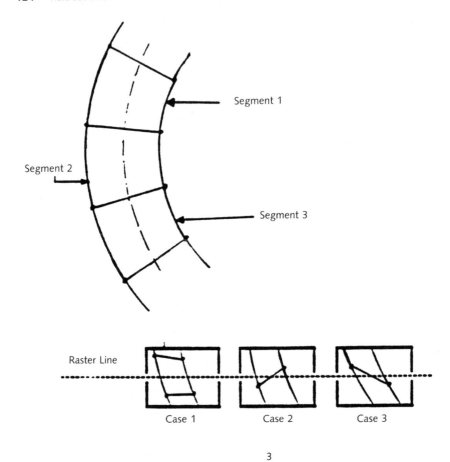

FIGURE 5.12.   Blending Band Case Definition.

line connecting the two intersected segments determines which case exists. When the slope is positive, case 2 applies. When the slope is negative, case 3 is applicable.

Case 1 is the simplest. It covers the situation where a raster line $y = Y_j$ intersects only one blending segment. The value of each pixel can be evaluated strictly on its position relative to the start and end of the blending region. In this case, the only remaining information to be determined are the x values of the outer and inner boundaries.

6.2.1.   The Average Circle.

To find the intersection of the raster line $y = Y_j$, where $Y_{min} < Y_j < Y_{max}$, with the outer boundary, the equation of the outer boundary is determined in a manner

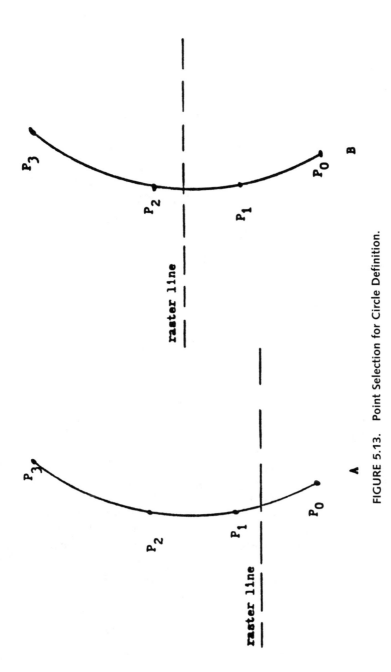

FIGURE 5.13. Point Selection for Circle Definition.

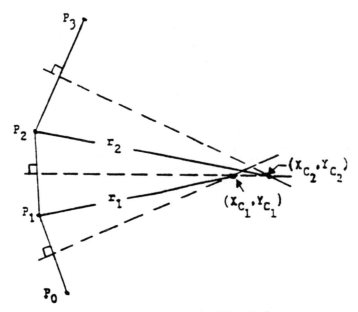

FIGURE 5.14.   Average of the Circles.

similar to that for determining (X_{max}, Y) from equation (1). The problem varies from the earlier description in that two cases may occur as shown in Figure 5.13. Figure 5.13A is analogous to the situation in Figure 5.11, where the raster line crosses the arc with endpoints P_0 and P_1. It is clear that P_2 will be used as the third point in determining the equation of the circle containing P_0 and P_1 because it is the only point adjacent to the arc.

The second possibility, shown in Figure 5.13B, has boundary points existing on either side of the intersected arc. Instead of randomly selecting three of the four points, a slightly different method was developed which has produced better results. The method utilizes all four points by constructing the equations of two circles, each using one of the additional boundary points. These circles can then be used to construct an "average" circle (see Figure 5.14). The radius of the circle is found by averaging the two radii r_1 and r_2. The center is found by averaging the x and y values from the circle centers already found:

$$(X_C, Y_C) = ((X_{C_1} + X_{C_2})/2, (Y_{C_1} + Y_{C_2})/2).$$

The intersection point of the current raster line $y = Y_j$ and the average circle as defined in equation (1) is (X_{O_j}, Y_j) where the value of X_{O_j} is obtained by substituting (X_{O_j}, Y_j) for (X_O, Y_O) in equation (1) and then solving for X_{O_j} to give:

$$X_{O_j} = X_C - [r^2 - (Y_j - Y_C)^2]^{1/2}.$$

The same procedure is used to find the intersection of the raster line with the inner boundary, yielding the point of intersection as (X_{I_j}, Y_j). Once (X_{O_j}) and (X_{I_j}, Y_j) are known, all pixels on the raster line $y = Y_j$, for $X_{O_j} < X < X_{I_j}$ can be filled using the raster fill scheme to be discussed in the next section.

6.2.2. Crossing Blending Regions.

As raster lines are evaluated from bottom to top it is clear that eventually a raster line $y = Y_j$ will intersect more than one blending segment, resulting in either case 2 or case 3 shown in Figure 5.12B. Let the outer and inner boundary points on the segment defining the separation of blending segments be (X^*_O, Y^*_O) and (X^*_I, Y^*_I), respectively. Any raster line $y = Y_j$, such that

$$Y^*_O > Y_j > Y^*_I \qquad\qquad \text{(Case 2)}$$

or

$$Y^*_O < Y_j < Y^*_I \qquad\qquad \text{(Case 3)}$$

will pass through more than one segment. In these cases, the point of intersection with the segment connecting the outer and inner boundary points and the raster line must also be determined.

The intersection of the raster line with the outer and inner boundary is found in the same manner as for case 1. The equation of the line formed by the inner and outer boundary points is:

$$y = Y^*_O + m(X - X^*_O) \qquad\qquad (2)$$

where

$$m = (Y^*_O - Y^*_I) / (X^*_O - X^*_I)$$

Substituting Y_j for y in equation (2) yields X_{mid}, as shown in Figure 5.15. This point is important in determining how the raster fill process is conducted for case 2 and case 3. Once a raster line has been evaluated and determined to satisfy the requirements for case 2 or case 3, each succeeding raster line, as evaluation proceeds toward Y_{max}, will be of the same case type until the line again intersects only one blending region. At this point, the requirements for case 1 will again be satisfied. For example, in Figure 5.15, case 2 applies for each raster line $y = Y_j$ until $Y_j > Y^*_I$, at which time case 1 will again apply.

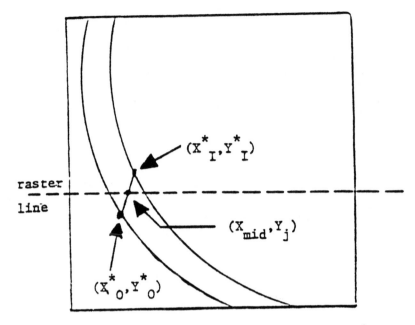

raster
line

FIGURE 5.15.   Raster Line Intersection With Segment Boundary.

7.   RASTER LINE FILLING

The raster line filling process uses the blending region information described in the previous sections to perform the final step of the blending process. This phase requires determination of how video memory must be modified to blend the two hemispherical displays together.

In Figure 5.16, a typical raster line and blending region segment are shown. All pixels outside the blending region can immediately be assigned a value of zero or one. The area to the left and/or below the blending region will have an intensity represented by zero. All pixels to the right and/or above the blending region will have an intensity represented by one.

7.1.   Raster Filling for Spans

Writing to video memory is always done on an element by element basis. However, when a constant value is to be written out for consecutive pixels located on the same raster line, the raster line filling process requires only the starting and ending locations of the span plus the single constant value to be applied to each pixel.

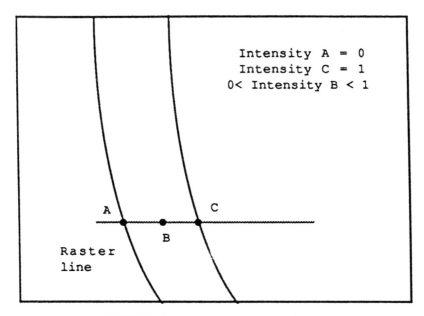

Intensity A = 0
Intensity C = 1
0< Intensity B < 1

A

C

B

Raster
line

FIGURE 5.16.   Blending Region Intensity.

Before writing to video memory, the span under consideration must be transformed from standard space to raster space. Let (X_S, Y_j) and (X_E, Y_j) be the respective starting and ending points for a span in standard space along the line $y = Y_j$. The inverse transformation matrix C_i^{-1} must be applied to each of these two points to transform them from standard space to raster space. The result of transforming these points back into raster space yields (X_{SR}, Y_{SR}) and (X_{ER}, Y_{ER}). When the starting and ending y values of the span in raster space are the same, $Y_{SR} = Y_{ER}$, the span is located on a single raster line and all values of video memory between X_{SR} and X_{ER} on the line $y = Y_{SR}$ are set to a constant using a poke process. In this case, there is no need to evaluate each element individually. However, when $Y_{SR} \neq Y_{ER}$, the span is not on one raster line and each pixel on the span in standard space must be transformed to raster space individually before a poke process can be performed to modify video memory.

7.2. Blending Region Filling

Blending will affect all elements on a raster line between the outer and inner boundaries of the blending region. The intensity of each element on the raster line will be ramped between one and zero using a linear ramping function. A linear function was chosen for these calculations because it yielded excellent results and it was quick to evaluate. Figure 5.17 shows the ramping function as

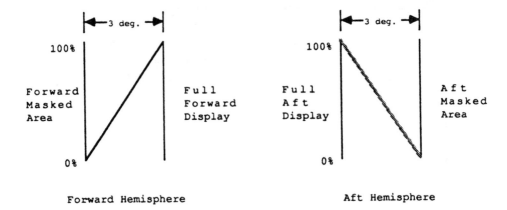

Forward Hemisphere   Aft Hemisphere

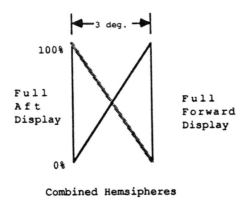

Combined Hemsipheres

FIGURE 5.17.   Video Intensity Ramp.

used on each hemispherical display, and on the two hemispherical displays combined.

The elements on a raster line within the blending region must be handled individually since no spans of constant intensity can be assumed to exist. Each pixel location along the raster line in standard space is transformed to raster space using the appropriate inverse transformation matrix C_i^{-1}. The value to be assigned is determined. Again the idea of three distinct cases is used in the raster fill process (see Figure 5.12B).

When case 1 exists, where the raster line intersects one blending segment only, each pixel on the raster line is assigned a value based on its position relative to the outer and inner boundary locations on the raster line. The x value for both the outer and inner boundaries on the raster line being evaluated are known as X_O

and X_I, respectively. The distance, D_O, from any point X where $X_O < X < X_I$, to the outer boundary, X_O, is found by

$$D_O = X - X_O.$$

Its distance from the inner boundary, D_I, is

$$D_I = X_I - X.$$

The value, K_p, to which the element being evaluated is to be set, can now be found by calculating:

$$K_p = K_{max} * D$$

where K_{max} is the value for maximum intensity and

$$D = D_O / D(D_O + D_I) \tag{3}$$

is a normalization factor. If a system is used which allows eight bits to determine intensity, $K_{max} = 255$.

When cases 2 or 3 exist, the point of intersection (X_{mid}, Y_j) with the raster line $y = Y_j$ (see Figure 5.15) is also used in the evaluation. This point allows the segment to be divided into two subsegments, the first between X_O and X_{mid} and the second between X_{mid} and X_I. Each subsegment can be evaluated separately in a manner similar to that described for use when case 1 conditions are satisfied if the intensity associated with the point (X_{mid}, Y_j) is known. This value, known as K_{mid} is found by computing:

$$K_{mid} = K_{max} * (X_{mid} - X_O) / (X_I - X_O)$$

which assigns the intensity at the intersection point as a percentage of the maximum intensity, set according to the location of intersection point with respect to the outer and inner boundaries.

The first subsegment, from X_O to X_{mid}, is filled in the same manner as for case 1 but substituting X_{mid} for X_I. This results in

$$D_O = X - X_O$$

and

$$D_I = X_{mid} - X$$

where $X_O < X \leq X_{mid}$. Using D_O and D_I, the normalization term D is found from equation (3). Similarly, when $X_{mid} < X < X_I$, X_{mid} is substituted for X_O so that

$$D_O = X - X_{mid}$$

and

$$D_I = X_I - X.$$

The associated intensity, K_p, can be found by computing:

$$K_p = K_{mid} * D \quad \text{when } X_O < X < X_{mid}$$

and

$$K_p = K_{max} * D \quad \text{when } X_{mid} \leq X < X_I.$$

8. CONCLUSIONS

The technique described for intensity blending of computer image generation-based displays was successfully implemented on Naval Training Device 2E6, Air Combat Maneuvering Simulator, NAS Oceana, by McDonnell Aircraft Company. This was the first implementation of the procedure. Using a General Electric Compu-Scene III computer image generation system, hemispherical background images were generated which were projected using two light valve projectors [9].

The linear ramping function originally used combined display intensities so they summed to 100 percent [10]. This function may also be useful for other applications. However, it is anticipated that given other applications, with other light valves or optics packages, a new offset ramp function might be needed. Further research suggested an offset linear function whose graph is shown in Figure 5.18. The function produced more pleasing results because of the response of the light valves. This response showed that, ideally, a smaller offset was needed for brighter scenes, and a larger offset was needed for dimmer scenes. The 160/255 offset shown in Figure 5.18 was a compromise of the above requirements that proved best for this particular application. Insufficient data is available to determine if this offset ramp will satisfy all similar applications or if it must be modified for each new application.

The blending algorithm can be used to control intensity effectively and realistically when joining hemispherical displays. Preliminary investigations indicate that it may be applicable to situations requiring nonhemispherical displays be projected onto the dome. For example, an area of interest display could be inserted as a patch into the forward display area, covering from 30 to 60 degrees. This provides the capability of displaying a specific area at a higher resolution which would allow much more detail to be apparent. For air to ground work, such as takeoffs and landings or bombing runs, this is an important tool. It is

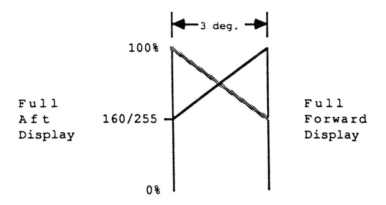

FIGURE 5.18.   Offset Ramp Function.

desirable to blend the forward display area with the inset such that the eye can perceive no abrupt change in the display. By modifying the blending algorithm to handle different shapes and revising the definition of standard space, it appears that the blending algorithm can be used to blend nonhemispherical displays.

REFERENCES

1. D.B. Coblitz, "Training Techniques Using Computer Generated Imagery (CGI)," *National Security Industrial Association Interservice/Industry Training Equipment 2nd Conference and Exhibition,* 1980, pp. 71–83.

2. W.B. Scott, "Visual Systems Refine Flight Simulation," *Aviation Week & Space Technology,* Vol. 119, No. 16, 1983, pp. 105–113.

3. "AV-8B Simulator Designed for Combat Missions," *Aviation Week & Space Technology,* Vol. 117, No. 18, 1982, pp. 86–89.

4. *Compu-Scene III, Operations and Maintenance,* Tech. Rep. MCAIR-2E6, St. Louis, MO, McDonnell Aircraft Co., 1986.

5. R.C. Hebb, *Visual Display Parameters,* Tech. Rep. NAVTRAEQUIPCEN IH-356, Naval Training Equipment Center, Simulation Technology Branch, 1984.

6. K.J. Stein, "USAF Evaluates New Simulator Imagery," *Aviation Week & Space Technology,* Vol. 120, No. 24, 1984, pp. 72–75.

7. *Proposal for Modification of F-14A Device 2F112 WAVS,* Tech. Rep. MDC B0517, Rev. A, Vol. IIA, St. Louis, MO, McDonnell Aircraft Co., 1987.

8. J.L. Bentz, "Joining Techniques for Optically Combined Visual Display Systems," *National Security Industrial Association/Industry Training Equipment 2nd Conference and Exhibition,* 1980, pp. 46–59.

9. K.J. Stein, "Simulation Techniques Converging to Meet Military/Commercial Needs," *Aviation Week & Space Technology,* Vol. 122, No. 11, 1985, pp. 239–243.

10. F.D. Eckelmann, *Device 2E6 Blending Circuit Notes.* St. Louis, MO: McDonnell Aircraft Co., 1986.

6
Long-Range Adaptive Control Algorithms for Robotics Applications*

J.M. Lemos
F. Coito
P. Shirley
P. Conceição
INESC, R. Alves Redol 9, 1000 Lisboa, Portugal

F. Garcia
C. Silvestre
J.S. Sentieiro
CAPS, Complexo I, Instituto Superior Técnico, Av. Rovisco Pais 1, 1096 Lisboa Codex-Portugal.

1. INTRODUCTION

The control of robot manipulators is an important application area of Adaptive Control. Since adaptive controllers are able to maintain good performance over a wide range of motions, payloads, and working conditions, they have attracted the attention of many researchers.

Several adaptive algorithms may be considered, of which not all are equally robust or present the same performance. The approaches to adaptive robot control can be formulated in continuous or discrete time, use SISO or MIMO techniques, linear or nonlinear models, and various design techniques [1, 2, 3, 4]. Model reference adaptive control (MRAC) is commonly used [5, 6, 7, 8, 9, 10]. Controller gains are adjusted by a minimization scheme (e.g., steepest descent) in order that a reference model for the closed-loop dynamics is followed. The simplicity of the method has the drawback that stability analysis is critical. Other approaches include the use of nonlinear techniques and feed-forward [11, 12, 13,

* Several of the ideas presented and much of the "momentum" to produce this work stem from Edoardo Mosca of the University of Florence, Italy, whom the authors wish to thank.

14, 15], self-turning algorithms [16, 17, 18], PD structures [19, 20], and the large family of long-range predictive controllers to be considered here [21, 22, 23, 24, 25, 26, 27, 28].

A significant number of these algorithms, however, turn out to be nonrobust with respect to unmodeled dynamics, requiring a number of artificial "fixes" in order to work. As an example, high-velocity arms using flexible structures for increased lightness may hardly be controlled with most of these strategies. Indeed, it has been shown [29] that unless proper precautions are taken, the presence of band-pass unmodeled dynamics can easily destroy the convergence and stability properties of adaptive algorithms.

The synthesis of self-tuning controllers based on long-range predictive models is receiving an increased interest. A number of algorithms, attempting to minimize a quadratic cost defined over an extended-time horizon, have been developed in order to overcome the limitations of standard self-tuning controllers, inherent to the adoption of a single-step ahead cost functional. Basically, such controllers aim at approximating the LQ stochastic control law over a semi-infinite horizon. MUSMAR [21], IMRHAC [30, 31], and GPC [32, 33] are relevant examples.

This work reviews long-range adaptive controllers in discrete-time based on the minimization of multistep ahead quadratic cost functionals. Several algorithms are derived in a unified basis. Stability and convergence aspects are discussed, together with several application examples to robotics, covering:

- Adaptive control of a flexible arm with reduced complexity controllers.
- Two-joint rigid manipulator, using SISO and MIMO algorithms and including both simulations and tests with a real arm.
- Depth and pitch control of an underwater remote-operated vehicle.

The text is organized as follows:

- Section 2 reviews the theory and the main implementation issues related to long-range adaptive control algorithms.
- Section 3 is devoted to applications of manipulator-motion control.
- Section 4 considers the control of an underwater ROV, and finally,
- Section 5 summarizes the conclusions to be taken as the "message" of this work.

2. LONG-RANGE ADAPTIVE CONTROL ALGORITHMS

This section reviews the theory and the main implementation issues related to long-range adaptive control algorithms. The control problems underlying long-range algorithms are explained first. The predictive modela, upon which the solution to these problems rely, are then studied (both things are then gathered in order to obtain adaptive controllers). For the sake of simplicity, discussion is done for the SISO case, and the MIMO case is briefly reviewed at the end.

2.1. Long-Range Control

In *long-range adaptive controllers* (Figure 6.1), the plant is described over a certain *time horizon* by a set of *predictive models*. At each time t, among the admissible control scenarios, the strategy which optimizes the given criterion is chosen. The resulting control is then applied to the plant, but *only* at time t. According to a *receding-horizon* strategy, the whole procedure is repeated in the next sampling period. Thus, instead of choosing the manipulated variable at time t, u_t, such that the output of the plant at time $t + 1$, y_{t+1} is some desired value, the evolution of the plant over more steps is taken into consideration.

The concept of a trajectory connecting the present output with the set-point at the end of the time horizon is important. This consists in assuming that, over the

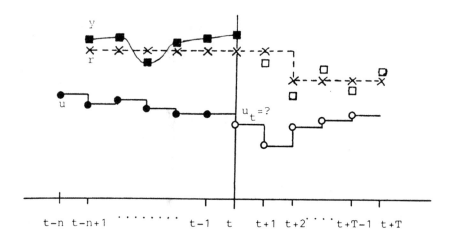

$\bullet$ – Observations of u, known up to time $t - 1$

$\square$ – Observations of y, known up to time t

$\times$ – Refererence to track

$\circ$ – Sequence of control samples, from time t up to $t + T - 1$

$\square$ – y forecasted, from $t + 1$ up to $t + T$

FIGURE 6.1.   Interpretation of Long-Range Adaptive Control

control horizon, the output should follow a "reference trajectory" w_i^*, starting at $y(t)$ and ending in $w(t + T)$, the set-point at the end of the horizon. A possible choice is

$$w_i^*(t) = \gamma w_{i-1}^*(t) + (1 - \gamma)w(t + T) \tag{1}$$

$$w_0^*(t) = y(t) \tag{2}$$

$$w_T^*(t) = w(t + T) \tag{3}$$

where $0 \leq \gamma < 1$ is a parameter.

Several types of models have been used in order to describe the plant behavior over the control horizon. Convolution with the impulse and step responses, transfer functions and linear-prediction models are common choices [34]. (These will be discussed further in the next subsection.)

In what concerns the criterion for control selection, one of the main possibilities is to choose the manipulated variable in order to minimize the *multistep quadratic cost functional*

$$J_T = \frac{1}{T} \epsilon \left\{ \sum_{k=1}^{T} [\bar{y}^2(t + k) + \rho u^2(t + k - 1)] | I^t \right\} \tag{4}$$

in which

$$\bar{y}(t + k) \triangleq y(t + k) - w_k^*(t) \tag{5}$$

is the tracking error with respect to the reference trajectory, ρ is a non-negative weight on the penalty of the control effort, and I^t is the information available up to time t.

If T is large enough, the steady-state solution of the LQ stochastic problem

$$\min_u \lim_{t \to \infty} \epsilon[\bar{y}_t^2 + \rho u_t^2] \tag{6}$$

is approximated. It is also expected that the controller is insensitive to the plant i/o transport delay.

For the minimization of (4), two main possibilities may be considered (Figure 6.2):

1. either assume the first N_u samples of the control, from time t up to time $t + N_u - 1$, to be free, and choose them in order to minimize J_T. The remaining samples, from time $t + N_u$ up to time $t + T - 1$ are constrained to be constant and equal to u_{t+N_u-1};
2. or assume the control samples between $t + 1$ and $t + T - 1$ to be given by a

GPC, 1MRHAC

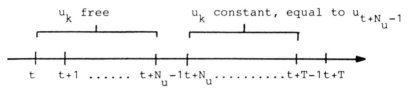

MUSMAR

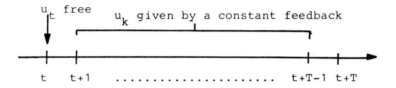

FIGURE 6.2.   Strategies for Long-Range Quadratic Control

constant feedback of a "state" of the plant, and choose u_t such that J_T is minimized.

Choice 1 leads to algorithms such IMRHAC [30, 31] and GPAC [32, 33]. When $N_u = T$ (i.e., when all the control samples over the horizon are free), the minimization process is equivalent to iterate backward T steps of the Riccati equation, in a state-space formulation of the LQ stochastic problem [35]. Clearly, since, of the calculated sequence, u_t is the only control actually applied to the plant, if T is large enough, the resulting closed-loop will be stable.

Choice 2, instead, leads to algorithms such as MUSMAR [21] and SPM [36]. For $T = \infty$, this optimizaiton procedure is equivalent to Kleinman's iterations to solve the algebraic Riccati iteration. This has the consequence that the convergence rate with T, of the calculated feedback to the steady-state LQ solution, is much faster in Choice 2 than in Choice 1. Besides a better control formulation, as will be further discussed, the imposition of a constant feedback improves the adaptation mechanism.

2.2.   Predictive Models

A robot manipulator is a highly *nonlinear* and intrinsically *multivariable* system. However, it can be modeled locally as a linear system *provided that its model is continually updated.* Furthermore, provided that interactions between joints are small, and that the controller to be used is able to deal with output loads, each joint can be controlled independently of the other. This approach has been followed (e.g., by [16]).

Thus, in order to derive the predictive models used for the minimization of (4), consider the SISO plant described by the ARX model

$$A(q)y(t) = B(q)u(t) + q^{\partial A}e(t) \tag{7}$$

in which A and B are polynomials in the forward-shift operator q, such that

$$\partial A - \partial B = d \geq 1 \tag{8}$$

∂A denotes the degree of A, A is monic, and the common factors between A and B are stable (i.e., all noncontrollable or nonobservable modes are stable). Hereafter, the *nominal* value of d is taken as one.

The innovations signal $\{e(t)\}$ is a white-noise sequence, such that

$$E[e(t)e(t + k)] = \sigma_e^2\delta_k \tag{9}$$

where δ_k is the kronecker symbol. Latter, the extension to the colored-noise case will be discussed.

Introducing the *recyprocal polynomial*

$$A^*(q^{-1}) \triangleq q^{\partial A}A(q^{-1}) \tag{10}$$

model (7) is written in terms of delayed variables:

$$A^*(q^{-1})y(t) = B^*(q^{-1})u(t - 1) + e(t) \tag{11}$$

where

$$A^*(q^{-1}) = 1 + \sum_{i=1}^{n} a_iq^{-i} \tag{12}$$

$$B^*(q^{-1}) = \sum_{i=0}^{m} b_iq^{-i} \tag{13}$$

$$n \triangleq \partial A$$
$$m \triangleq \partial B.$$

Note that no hypothesis is done on the exact value of the i/o transport delay, and, thus, the first coefficients of the B polynomial are possibly null.

2.2.1. *Unconstrained Predictive Models.*

The theory of diophantine equations guarantees that there exists polynomials F_j and G_j of the form

$$F_j^*(q^{-1}) = 1 + f_1 q^{-1} + \ldots + f_j q^{-j} \tag{14}$$

$$G_j^*(q^{-1}) = g_0^j + g_1^j q^{-1} + \ldots + g_{n-1}^j q^{-n+1} \tag{15}$$

that satisfy the identity

$$1 = F_j^*(q^{-1})A^*(q^{-1}) + q^{-j-1}G_j^*(q^{-1}) \tag{16}$$

According to a standard argument, in order to obtain the $j + 1$ steps-ahead predictor for y, multiply (11) by $F_j^*(q^{-1})$ and use (16) to get

$$y(t + j + 1) = G_j^*(q^{-1})y(t) + F_j^*(q^{-1})B^*(q^{-1})u(t + j) + \\ + F_j^*(q^{-1})e(t + j + 1) \tag{17}$$

Since the last term of the second member of (17) is orthogonal to the others, the optimal estimator (in the sense that it minimizes the mean-quadratic error) of $y(t + j + 1)$, given information up to time t, is given by

$$\hat{y}(t + j + 1|t) = G_j^*(q^{-1})y(t) + F_j^*(q^{-1})B^*(q^{-1})u(t + j) \tag{18}$$

Define polynomials $E_j^*(q^{-1})$ and $H_j^*(q^{-1})$ by

$$F_j^*(q^{-1})B^*(q^{-1}) = H_j^*(q^{-1}) + q^{-j-1}E_j^*(q^{-1}) \tag{19}$$

$$H_j^*(q^{-1}) = h_1 + h_2 q^{-1} + \ldots + h_{j+1} q^{-j} \tag{20}$$

$$E_j^*(q^{-1}) = \zeta_0^j + \zeta_1^j q^{-1} + \ldots + \zeta_{m-1}^j q^{-m+1} \tag{21}$$

where, by (16), it is concluded that the coefficients of $H_j^*(q^{-1})$ are the first $j + 1$ samples of the impulse response of B/A.

The predictor (18) is written in the equivalent form

$$\hat{y}(t + j + 1|t) = G_j^*(q^{-1}) + H_j^*(q^{-1})u(t + j) + E_j^*(q^{-1})u(t - 1) \tag{22}$$

Defining the vector of coefficients

$$\pi_{j+1} \triangleq [g_0^j g_1^j \ldots g_{n-1}^j \zeta_0^j \zeta_1^j \ldots \zeta_{m-1}^j]' \tag{23}$$

and the *pseudostate*

$$s_t \triangleq [y_t y_{t-1} \cdots y_{t-n+1} u_{t-1} \cdots u_{t-m}]' \tag{24}$$

the following predictive model is obtained:

$$y(t + j + 1) = \sum_{i=0}^{j} h_{i+1} u_{t+j-1} + \pi'_{j+1} s_t + \epsilon_j(t) \tag{25}$$

where $\epsilon_j(t)$ is a residuo orthogonal to the other therms.

The above model (25) may be interpreted in the following way: The output at time $t + j + 1$ depends only on the psuedostate at time t and the controls applied between time t and time $t + j$ (apart from orthogonal terms). If $u_t = \ldots = u_{t+j} = 0$, only the free-response term $\pi'_{j+1} s_t$ is used in the prediction. If this sequence of future controls is non-null, the convolution summation must be added to the free-response term.

The set of adaptive control algorithms considered are based on sets of predictive models over a time horizon. So, model (25) is written down for $j = 0$ up to $T - 1$:

$$y(t + 1) = h_1 u_t + \pi'_1 s_t + \epsilon_1(t)$$
$$y(t + 2) = h_1 u_{t+1} + h_2 u_t + \pi'_2 s_t + \epsilon_2(t)$$
$$\cdots \cdots \cdots \cdots$$
$$y(t + T) = h_1 u_{t+T-1} + \ldots + h_T u_t + \pi'_T s_t + \epsilon_T(t)$$

Define

$$Y_t \triangleq [y_{t+1} \cdots y_{t+T}]' \tag{26}$$

$$U_t \triangleq [u_t \cdots u_{t+T-1}]' \tag{27}$$

$$\epsilon_t \triangleq [\epsilon_1(t) \cdots \epsilon_T(t)]' \tag{28}$$

$$H \triangleq \begin{bmatrix} h_1 & 0 & \cdots & 0 & 0 \\ h_2 & h_1 & \cdots & 0 & 0 \\ \cdots & \cdots & \cdots & \cdots & \cdots \\ h_T & h_{T-1} & \cdots & h_2 & h_1 \end{bmatrix} \tag{29}$$

$$\Pi \triangleq [\pi_1 \cdots \pi_T] \tag{30}$$

The set of predictive models is written

$$Y_t = H U_t + \Pi' s_t + \epsilon_t \tag{31}$$

This *multipredictor model* will be used for the derivation of adaptive control algorithms (IMRHAC and GPC).

2.2.2. MUSMAR Models.

As was already referred, the MUSMAR algorithm restricts the future control samples, from time $t + 1$ up to $t + T - 1$, to be given by a *fixed feedback* of the pseudostate, leaving only u_t free. In order to see how the predictive model (31) is modified by this assumption, start by observing that the pseudostate s_t satisfies the dynamic equation

$$s_{t+1} = \Phi_s s_t + \Gamma u_t + \bar{e}_1 e_t \tag{32}$$

in which

$$\bar{e}_1 \triangleq [1\ 0\ \ldots\ 0]' \tag{33}$$

$$\Phi_s \triangleq \begin{bmatrix} \psi_1' \\ I_{n-1}\ \underline{0}_{(n-1)\times n} \\ \underline{0}_{1\times(n+m)} \\ \underline{0}_{(m-1)\times n}\ I_{m-1}\ \underline{0}_{(m-1)\times 1} \end{bmatrix} \tag{34}$$

$$\Gamma = [b_0\ 0\ \ldots\ 0\ 1\ 0\ \ldots\ 0]' \tag{35}$$

$$\psi_1' = [-a_1\ \ldots\ -a_n\ b_1\ \ldots\ b_m] \tag{36}$$

Assume that a constant feedback law of the type

$$u_{t+k} + F_0' s_{t+k} + \eta_{t+k} \tag{37}$$

is acting on the plant from time $t + 1$ up to time $t + T - 1$. The signal $\{\eta_t\}$ is a white-noise dither sequence uncorrelated with $\{e_t\}$ and such that

$$\sigma_\eta^2 \triangleq E[\eta_t^2] \ll \sigma_e^2$$

Thus, from time $t + 1$ up to $t + T - 1$, the pseudostate verifies

$$s_{t+k+1} = \Phi_F s_{t+k} + \Gamma \eta_{t+k} + \bar{e}_1 e_{t+k} \tag{38}$$

with

$$\Phi_F \triangleq \Phi_s + \Gamma F_0' \tag{39}$$

Combining (32) and (38), it is concluded that the optimal predictor of s_{t+i} given information up to time t is

$$\hat{s}_{t+i|t} = \Phi_F^{i-1}[\Phi_s s_t + \Gamma u_t] \tag{40}$$

The optimal predictor for u_{t+i} is

$$\hat{u}_{t+i|t} = F'_0 \hat{s}_{t+i|t} \tag{41}$$

i.e.,

$$\hat{u}_{t+i|t} + F'_0 \Phi_F^{i-1}[\Phi_s s_t + \Gamma u_t] \tag{42}$$

Define

$$\mu_i \triangleq f'_0 \Phi_F^{i-1} \Gamma \tag{43}$$

$$\phi'_i \triangleq f'_0 \Phi_F^{i-1} \Phi_s \tag{44}$$

The following expression for the optimal predictor of u_{t+i} is obtained

$$u_{t+i} + \mu_i u_t + \phi'_i s_t + v^u_i(t) \tag{45}$$

where $v^u_i(t)$ is a residuo orthogonal to the other therms. Also

$$\hat{y}_{t+i|t} = [1\ 0\ \ldots\ 0]\hat{s}_{t+i|t} \tag{46}$$

i.e.,

$$\hat{y}_{t+i|t} = \bar{e}'_1 \Phi_F^{i-1}[\Phi_s s_t + \Gamma u_t] \tag{47}$$

Define

$$\theta_i \triangleq \bar{e}'_1 \Phi_F^{i-1} \Gamma = (\psi'_1 + b_0 F'_0) \Phi_F^{i-2} \Gamma \tag{48}$$

$$\psi'_i \triangleq \bar{e}'_1 \Phi_F^{i-1} \Phi_s = (\Psi'_1 + b_0 F'_0) \Phi_F^{i-2} \Phi_s \tag{49}$$

The following expression for the optimal predictor of y_{t+i} is, thus, obtained

$$y_{t+i} = \theta_i u_t + \psi'_i s_t + v^y_i(t) \tag{50}$$

The considerations above prove the following:
Proposition 1.
The ARX plant (7) working under a constant feedback control from time $t + 1$ up to $t + T - 1$, admits predictive models of the form

$$Y_t = \Theta u_t + \Psi' s_t + v^y(t) \tag{51}$$

$$U_t = \mu u_t + \Phi' s_t + v^u(t) \tag{52}$$

where

$$\Theta \triangleq [\theta_1 \ . \ . \ . \ \theta_T]' \tag{53}$$

$$\Psi \triangleq [\psi_1 \ . \ . \ . \ \psi_T] \tag{54}$$

$$\mu \triangleq [1\mu_1 \ . \ . \ . \ \mu_{T-1}]' \tag{55}$$

$$\Phi \triangleq [\underline{0}\phi_1 \ . \ . \ . \ \phi_{T-1}] \tag{56}$$

$\square$

It is remarked that, unlike in the multipredictor model (31), the parameters of (51, 52) depend on the feedback applied to the plant. Only the one-step ahead predictor depends exclusively on the plant, being given by

$$\theta_1 = b_0 \tag{57}$$

$$\psi_1 = [-a_1 \ . \ . \ . \ -a_n \ b_1 \ . \ . \ . \ b_m]' \tag{58}$$

The coefficients θ_i and μ_i admit an interpretation given by the following:
Proposition 2.

The parameters θ_i and μ_i are, respectively, the first T samples of the impulse response of the system whose input is η and whose output is y (for the θ_i), and u (for the μ_i)

$$\sum_{i=1}^{T} \theta_i q^{-i} = \left. \frac{q^{-k}B^*(q^{-1})}{Q^*(q^{-1})} \right|_T \tag{59}$$

$$\sum_{i=0}^{T-1} \theta_i q^{-i} = \left. \frac{A^*(q^{-1})}{Q^*(q^{-1})} \right|_T \tag{60}$$

where Q is the closed-loop characteristic polynomial, and $H|_T$ means the truncation of order T of the series expansion in q^{-1} of H. $\square$

$\square$

Before proving *Proposition 2*, the following Lemma is stated, whose proof is to be found in [38]:
Lemma 1.

Let the transfer function between u and y admit the state-space realization

$$x_{t+1} = \Phi x_t + Gu_t \tag{61}$$

$$y_t = Hx_t \tag{62}$$

Then, the samples of its impulse response are given by

$$h_i = H\Phi^{i-1}G \tag{63}$$

□

Proof of Proposition 2.

The transfer functions between the dither, the output, and input are given, respectively, by

$$y_t = \frac{q^{-k}B^*(q^{-1})}{Q^*(q^{-1})}\eta_t \tag{64}$$

$$u_t = \frac{A^*(q^{-1})}{Q^*(q^{-1})}\eta_t \tag{65}$$

where Q is the closed-loop characteristic polynomial.

Consider (64). From the discussion preceding Proposition 1, it is clear that this transfer function admits the state-space realization

$$s_{t+1} = \Phi_F s_t + \Gamma\eta_t \tag{66}$$

$$y_t = \bar{e}'_1 s_t \tag{67}$$

where $\bar{e}'_1 = [1\,0\,\ldots\,0]$.

By Lemma 1, the samples of the impulse response of (64), in terms of the state-space realization (66, 67), are given by

$$h_i^y = \bar{e}'_1\Phi_F^{i-1}\Gamma \tag{68}$$

i.e., they are equal to θ_i (compare with (48)).

Consider now (65). Using the same arguments as before, observe that there is a state-space realization given by (66) and

$$u_t = [0\,\ldots\,0\,1\,0\,\ldots\,0]s_{t+1}$$
$$u_t = [0\,\ldots\,0\,1\,0\,\ldots\,0]\phi_F st + [0\,\ldots\,0\,1\,0\,\ldots\,0\Gamma\eta_t$$
$$u_t = F'_0 s_t + \eta_t \tag{69}$$

The samples of the impulse response of (65) are, thus,

$$h_i^u = F'_0\Phi_F^{i-1}\Gamma \tag{70}$$

i.e., the μ_i (compare with (43)).

□

2.2.3. The Coloured-Noise Case.

The derivation of models (31) and (51, 52) is much more involved in the coloured-noise case. It suffices, here, to say that an ARMAX plant admits in closed-loop predictive models of the form (31) with the residuo orthogonal to the data in the regressor, if its control is generated by a feedback of an optimal estimate of a state. Assuming a constant feedback, MUSMAR predictive models of the form (51, 52) can be obtained. This is the subject of the *Implicit Modeling Theory*, details of which can be found in [31, 36, 37].

The relevance of this issue stems from the fact that even plants with coloured noise admit predictive models which can be identified by standard-recursive least squares (RLS), which are more robust than generalized, or extended, RLS.

To conclude, it can be said that the *ARMAX plant with accessible disturbance v(t)*

$$A(q)y(t) = B(q)u(t) + D(q)v(t) + C(q)e(t) \tag{71}$$

admits in closed-loop, under mild conditions, predictive models of the form (31) and (51, 52), with s_t a vector of *finite* length composed of samples of the output, the input, the reference, and the accessible disturbance, such that the residues are orthogonal to the data (but not mutually independent).

2.2.4. *Parameter Estimation.*

In what concerns parameter estimation for the multipredictor models obtained in the previous sections, there are two main problems to be considered:

1. the reduction of the *computational load* associated with the estimation of sets of predictors (*T* in one case and *2T* in the other). Most of the data upon which the predictors rely is common to all of them, a fact exploited in order to obtain efficient algorithms.

2. *parameter identifiability.* Since the gain of the plant is estimated in closed-loop, the dither signal {η} is introduced in order to make all the parameters identifiable. For obvious reasons related with control performance, the variance of the dither must be kept small or even vanish. The parameters will then tend to drift, and this may be a source of trouble. Thus, a mechanism must be incorporated with "freezes" the parameter estimates when no new information is arriving on them. Algorithms, such as the *variable forgetting factor* [39], are ineffective since they are not able to discriminate between the various parameters, with various degrees of identifiability. Instead, with the *directional forgetting* version of recursive least squares (DFRLS) [40, 41, 42], the estimates of identifiable parameters are improved while the estimates of unidentifiable ones are frozen.

   2.2.4.1. *Reducing the Computational Load.* Parameter estimation of models like

$$Y_t \approx \Theta u_t + \Psi' s_t \tag{72}$$

$$U_t \approx \mu u_t + \Phi' s_t \tag{73}$$

where $\approx$ denotes equality in least-squares sense, involve a very small computational burden since *all the predictors share the same regressor*, namely

$$\varphi(t) = [u(t)s'(t)]' \tag{74}$$

Thus, *only one* covariance matrix must be updated in a recursive least-squares (RLS) scheme and only the prediction errors must be individually calculated for each predictor. This leads to adaptive control algorithms with a small computational load, such as MUSMAR and SPM. The only possible drawback is that the first predictor (i.e., in general, the one-step ahead predictor) do not use the most recent data. This, however, has never appeared to be inconvenient in practice.

Parameter estimation in models of the form

$$Y_t \approx HU_t + \Pi' s_t \tag{75}$$

instead, causes additional problems since, due to the lower triangular structure of matrix H, on going from predictor j-steps ahead to $j + 1$, an additional sample of the input is needed in the data, and the regressor is no longer invariant. Several strategies are considered to reduce the computational load in this case.

Analytic extrapolation.

One possibility to estimate the parameters in (75) is to admit that the j-steps ahead predictor may be obtained from the one-step ahead predictor by analytic extrapolation. This may be done using Algorithm 1, being the approach followed in the GPC algorithm. In order that the standard RLS algorithm may be used to estimate the parameters of the one-step ahead predictor, it is assumed that the plant is ARX. This leads to an error in the estimates when the plant is ARMAX. Other sources of error are the fact that the plant is nonlinear, or unmodelled dynamics are present.

Algorithm 1 [32].
In order to get estimates of the parameters of the multipredictive model

$$y_{t+j+1} \approx \sum_{i=0}^{j} h_{i+1} u_{t+j-i} + \Pi'_{j+1} s_t \tag{76}$$

$$j = 0, \dots, T - 1$$

$$s_t = [y_t \dots y_{t-n} u_{t-1} \dots u_{t-m}]' \tag{77}$$

using an analytic extrapolation procedure, execute the following steps:

1. Using RLS estimate the parameters in the one-step ahead predictor

$$y_{t+1} \approx h_1 u_t + \Pi_1' s_t \tag{78}$$

2. Let

$$a_i = -\Pi_1(i) \qquad i = 1, \ldots, n$$
$$b_0 = h_1$$
$$b_i = \Pi_1(i + n) \qquad i = 1, \ldots, m$$

3. Let

$$f_0 = 1$$
$$g_i^0 = -a_{i+1} \qquad i = 0, \ldots, n - 1$$

4. For $j = 1$ up to $T - 1$ let

$$f_j = g_0^{j-1}$$
$$g_i^j = g_{i+1}^{j-1} - a_{i+1} f_j \qquad i = 0, \ldots, n - 2$$
$$g_{n-1}^j = -a_n f_j$$
$$F_j^*(q^{-1}) = 1 + \sum_{i=1}^{j} f_i q^{-i}$$
$$B^*(q^{-1}) = \sum_{i=0}^{m} b_i q^{-i}$$
$$\beta^j(q^{-1}) = F_j^*(q^{-1}) B^*(q^{-1})$$
$$h_{j+1} = \beta_j^j$$
$$\Pi_{j+1}(i) = g_{i-1}^j \qquad i = 1, \ldots, n$$
$$\Pi_{j+1}(n + i) = \beta_{j+i}^j \qquad i = 1, \ldots, m$$

$\square$

Interlaced estimation.

When the ARMAX model used to represent the plant is just an approximation due to the causes mentioned, the *separate estimation* of the predictors is a possible way to minimize the error. Also, the Implicit Modeling Theory implies that the predictors should be separately estimated when the noise is coloured.

One possibility to estimate the predictors separately is the following *interlaced scheme:*

Algorithm 2 [31].
In order to get estimates of the parameters of the multipredictive model (76) using an interlaced scheme, execute the following steps:

1. Let

$$z_{t+1} \triangleq y_{t+1}$$

2. Estimate w_1 and Π_1 from the model

$$z_{t+1} \approx w_1 u_t + \Pi'_1 s_t$$

3. For $j = 2$ up to T do
    a) Calculate

$$z_{t+j} \triangleq y_{t+j} - \sum_{i=1}^{j-1} w_i u_{t+j-i}$$

with the w_i replaced by their current estimates.
    b) Estimate w_j and Π_j from

$$z_{t+j} \approx w_j u_t + \Pi'_j s_t \qquad \qquad \square$$

Other methods.
Other methods for parameter estimation in the above predictive structures are obtained in [30, 36] as specializations of Bierman's UD algorithm [43] and the Square Root Information Filter [43].

2.2.4.2.   *Parameter Identifiability and Directional Forgetting RLS.* In the exponential forgetting version of RLS, the information matrix Λ is updated according to

$$\Lambda(t) = \Lambda(t-1) - (1-\lambda)\Lambda(t-1) + \varphi(t-1)\varphi'(t-1) \qquad (79)$$

where φ is the vector of data and λ is the forgetting factor. Hence, there is a (negative) "forgetting" term proportional to $\Lambda(t-1)$ (i.e., the forgetting is made *equally* along all the directions in the space of parameters). Suppose a combination of parameters or parameter defining a manifold $\mathcal{M}$ is unidentifiable. Due to the fact that information is equally lost in all space directions and no new information comes along $\mathcal{M}$, the diameter of ellipses of constant probability of parameter error will grow along $\mathcal{M}$ (Figure 6.3).

$\square$

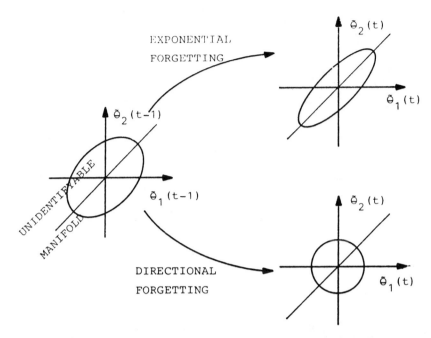

FIGURE 6.3. The Action of Exponential Forgetting and Directional Forgetting on the Ellipses of Constant Probability of Parameter Estimation Error, According to the Probabilistic Interpretation Given in [101]

Instead, in *directional forgetting* RLS [40, 41, 42] (DFRLS), the information matrix is updated according to

$$\Lambda(t) = \Lambda(t - 1) + (1 - \alpha(t))\varphi(t - 1)\varphi'(t - 1) \tag{80}$$

where $\alpha(t)$ is a scalar defining the amount of information being lost in the "direction" of the matrix of characteristic $1\varphi'(t - 1)\varphi(t - 1)$. Thus, information is deleted only in the "direction" of new arriving information, and unidentifiable parameters are "frozen." It is shown in [42] that, in terms of the probability density of the parameter-estimation error, this amounts to the equalization of the diameters of the ellipses of constant probability (Figure 6.3).

The equations of DRFLS for the estimation of the parameters θ in the predictive model

$$y(t) = \theta'\varphi(t - 1) + v(t) \tag{81}$$

in which y and φ are observables and v is a residuo, are

$$\epsilon(t) = y(t) - \hat{\theta}'(t-1)\varphi(t-1) \tag{82}$$

$$K(t) = \frac{P(t-1)\varphi(t-1)}{1 + \varphi'(t-1)P(t-1)\varphi(t-1)(1 - \alpha(t)]} \tag{83}$$

$$\hat{\theta}(t) = \hat{\theta}(t-1) + K(t)\epsilon(t) \tag{84}$$

$$P(t) = [I - K(t)\varphi'(t-1)(1 - \alpha(t))]P(t-1) \tag{85}$$

with

$$P(t) = A^{-1}(t) \tag{86}$$

One point to consider is the choice of the amount of information to forget, $\alpha(t)$. Clearly, $\alpha(t)$ must be non-negative, since that would mean an addition of information, instead of a removal. Also, from Equation (80), it is clear that the covariance matrix P may loose its positive definite character if $\alpha(t)$ becomes too big.

The following theorem, proved in [40], gives bounds on $\alpha(t)$, such that the P matrix is kept positive definite.

Theorem 1.

Let $P(t)$ satisfy (85). If $P(0) > 0$ then $P(t) > 0 \ \forall t$ if

$$0 \le \alpha(t) < 1 + \frac{1}{\rho(t)} \tag{87}$$

with

$$\rho(t) \triangleq \varphi'(t-1)P(t-1)\varphi(t-1) \tag{88}$$

☐

One possibility is to choose $\alpha(t)$, such that the P matrix approaches $\alpha_0 I$ with α_0 a scalar and I the identity matrix. This is the approach followed in [40]. However, it has shown in practice to be of little use and leads to an unnecessary computational burden.

Another possibility is to make

$$\alpha(t) = 1 - \lambda + \frac{1 - \lambda}{\varphi'(t-1)P(t-1)\varphi(t-1)} \tag{89}$$

$$0 < \lambda \le 1 \tag{90}$$

where λ may be seen as a forgetting factor in the "direction" of arriving information. This is equivalent to the approach of [41].

In practice, the covariance matrix P is propagated in factorized form [40, 41]. The implementation of factorized DFRLS using systolic arrays is studied in [44].

2.3. Adaptive Control Algorithms

The adaptive control algorithms to be considered are obtained coupling the long-range control laws minimizing the multistep quadratic cost functional (4) according to strategies one and two, with the estimation algorithms discussed in 2.2 according to the *Certainty Equivalence Principle* [45].

Thus, these main possibilities can be considered:

1. Assume strategy one (i.e., the first N_u samples of u over the control horizon are left *free*).
    *   Separately estimate the predictors in (31). This leads to IMRHAC [30, 31].
    *   Estimate the one-step ahead predictor and obtain future predictors by extrapolation. This leads to GPC [32, 33].
2. Assume strategy two (i.e., all the control samples over the optimization horizon are given by a *constant feedback* of the pseudostate).
    *   Separately estimate the predictors. This leads to MUSMAR [21].
    *   Estimate the one-step ahead predictor and obtain future predictors by extrapolation. This leads to SPM [36].

According to this, the detailed algorithms are:

Algorithm 3 (positional IMRHAC).

At each sampling period, recursively execute the following steps:

1. Given all i/o data up to time t, and using the *interlaced algorithm one*, recursively compute the DFRLS estimate of the parameters h_i and π_i in the set of predictive models

$$\bar{y}(t + i - T) = \sum_{j=1}^{i} h_j u_{t+i-T-j} + \pi_j' s_{t-T} + v^i(t) \qquad (91)$$

in which

$$s_t = [\bar{y}_t \ldots \bar{y}_{t-n_a+1} \, u_{t-1} \ldots u_{t-n_b} \, v_t \ldots v_{t-n_v+1}$$
$$w_{t+T-1} \ldots w_{t+T-n_w}]' \qquad (92)$$

$$\bar{y}_t \triangleq y_{t+i-T} - w_i^*(t) \qquad (93)$$

$v^i(t)$ is a residuo orthogonal to the data and the reference trajectory over the control horizon is calculated by (1, 2, 3).

2. Calculate the controller gains by

$$F = -\Pi H M^{-1} \bar{e}_1 \qquad (94)$$

with H a matrix given by (29), Π a matrix given by (30),

$$\bar{e}_1 \triangleq [1 \ 0 \ \ldots \ 0]' \tag{95}$$

and

$$M \triangleq \rho I + H'H \tag{96}$$

3. At sampling period t apply an input to the plant given by

$$u_t = F's_t + \eta_t \tag{97}$$

in which $\{\eta\}$ is a white-noise dither sequence, independent of the innovations of the plant. □

Algorithm 4 (positional GPC).
Equal to Algorithm 3, but instead of using the interlaced Algorithm 1 to perform the multipredictor parameter estimation, use the *extrapolation algorithm 2*. □

Remarks.

1. The above formulation assumes $N_u = T$. The algorithms with $N_u < T$ are formally the same, but with different matrices Π and H.
2. The vector of control inputs given by

$$U_t = \begin{bmatrix} u_t \\ \ldots \\ u_{t+T-1} \end{bmatrix} = -M^{-1}H\Pi s_t \tag{98}$$

minimizes the multistep quadratic cost (4). According to a receding-horizon strategy, only the first element of this sequence is actually applied to the plant, the whole procedure being repeated at the next sampling period. This leads to the control law (97), the dither η_t, being injected in order to fulfill a persistency of excitation condition, such that the parameters of the predictors become identifiable.

Algorithm 5 (positional MUSMAR).
At each sampling period, recursively execute the following steps:

1. Given all *i/o* data up to time t, recursively compute the DFRLS estimate of the parameters θ_i, ψ_i, μ_i, and ϕ_i in the set of predictive models:

$$\bar{y}_{t+i-T} = \theta_i u_{t-T} + \Psi'_i s_{t-T} + v^y_i(t-T) \tag{99}$$

$$u_{t+i-T-1} = \mu_{i-1} u_{t-T} + \Phi'_{i-1} s_{t-T} + v^u_{i-1}(t-T) \tag{100}$$

$$i = 1, \dots, T$$

$$v_i(t), \ v_i(t) \perp [u_t \ s_t]$$

2. Update the feedback parameters by

$$F = -\frac{\sum_{i=1}^{T} \theta_i \Psi_i + \sum_{i=1}^{T-1} \rho \mu_i \Phi_i}{\sum_{i=1}^{T} \theta_i^2 + \rho(1 + \sum_{i=1}^{T-1} \mu_i^2)} \tag{101}$$

and calculate the control by

$$u_t = F' s_t + \eta_t \tag{102}$$

where $\{\eta_t\}$ is a white-noise dither signal of small variance, independent of $\{e_t\}$. $\square$

Algorithm 6 (positional SPM).
Equal to Algorithm 5, but instead of estimating directly all the parameters in (99, 100), just identify

$$\bar{y}_t = \theta_1 u_{t-1} + \psi'_1 s_{t-1} + v^y_1(t-1) \tag{103}$$

and obtain the parameters for the future predictors using (43), (44), (48), and (49) and assuming that the current vector of controller gains will be kept constant.

1. The computational burden grows very little with the control horizon T. This is due to the fact that all the predictors in (99, 100) share a common regressor and, thus, *only one set of Kalman gains must be updated* in the RLS algorithm.
2. The predictive models for the output are not essential, since they can alternatively be derived from $\hat{\theta}_i$, $\hat{\Psi}_i$ and knowledge of previous feedback vectors.
3. For the first input predictor it is $\mu_0 = 1$ and $\Phi_0 = 0$, a fact already explored in Equation (101).
4. For $T = 1$ the classical self-tuning controller of Clarke and Gawthrop [46] is recovered, for plants with delay one.
5. MUSMAR is an approximation to an adaptive observer/controller for which the Kleinmans iterations of the Riccati equation are spread in time. This amounts to approximately minimize an unconditional stochastic quadratic-cost function by performing a spectral factorization. The above approximation is exact if $T = \infty$.

6. The robustness of MUSMAR with respect to plant i/o delay
    variations and/or uncertainty is due to the fact that it can be
    seen as a bank of T parallel self-tuners, each one matched to
    $E[\hat{y}^2((t + i + 1)|t) + \rho\hat{u}^2((t + i)|t)]$ and weighted by a
    probability distribution. This fact is established in [21] by
    noting that Equation (101) may be written as

$$F = \sum_i \hat{\Pi}_i \hat{F}_i$$

with

$$\hat{\Pi} = \frac{\hat{\theta}_i^2 + \rho\hat{\mu}_i^2}{\Sigma_{j=1}^T \hat{\theta}_j^2 + \rho\hat{\mu}_j^2}$$

$$\hat{F}_i = -\frac{\hat{\theta}_i \hat{\Psi}_i + \rho\hat{\mu}_i \hat{\Phi}_i}{\hat{\theta}_i^2 + \rho\hat{\mu}_i^2}$$

2.4.   Incremental Algorithms

In order to develop an incremental version of MUSMAR, consider the plant
model

$$A(q^{-1})y_t = B(q^{-1})u_{t-1} + d + v_t \tag{104}$$

in which A and B are polynomials in the unit-delay operator, satisfying conditions similar to the ones of model (91); d is an unacessible constant disturbance
and the stochastic process v_t satisfies

$$\Delta v_t = e_t \tag{105}$$

with

$$\Delta \triangleq 1 - q^{-1}$$

and e_t is as in model (91).
    Let $F_i(q^{-1})$ and $G_i(q^{-1})$ be the solutions of the following diophantine equation

$$1 - q^{-i} = \Delta F_i(q^{-1})A(q^{-1}) + \Delta q^{-i}G_i(q^{-1}) \tag{106}$$

Note that a solution always exists since the maximum common divisor of $\Delta A(q^{-1})$
and Δq^{-i} divides $1 - q^{-i}$.

Multiply (104) by $\Delta F_i(q^{-1})$ and use (106) to get the i-th steps-ahead predictive model

$$y_{t+i} = y_t + G_i(q^{-1})\Delta y_t + F_i(q^{-1})B(q^{-1})\Delta u_{t+i-1} + F_i(q^{-1})e_{t+i} \qquad (107)$$

which may be written

$$\bar{y}_{t+i} = \Pi_i' s_t + \sum_{j=1}^{i} h_j \Delta u_{t+i-j} + \epsilon_t^i \qquad (108)$$

where

$$s_t \triangleq [y_t \ \Delta y_t \ldots \Delta y_{t-n+1} \ \Delta u_{t-1} \ldots \Delta u_{t-m} \ w_{t+T}]' \qquad (109)$$

and

$$\epsilon_t^i \triangleq F_i(q^{-1})e_{t+i}$$

is a residuo orthogonal to the data, the vector Π_i has components depending on the polynomials $G_i(q^{-1})$ and $F_i(q^{-1})$. $B(q^{-1})$, and the h_j are samples of the step response of $B(q^{-1})/A(q^{-1})$.

In order to obtain the incremental version of MUSMAR, make the hypothesis that the *future values of the control variable are given by the constant feedback* F of the pseudostate, over the control horizon

$$\Delta u_{t+T-i} = F' s_{t+T-i} \qquad (110)$$
$$i = 1, \ldots, T-1$$

Under this hypothesis, and paralleling the analysis of 2.2.1., there exist matrices Φ_i and Γ_i, such that

$$s_{t+T-i} \approx \Phi_i s_t + \Gamma_i \Delta u_t \qquad (111)$$

where $\approx$ means equality in least-squares sense.

Coupling (108, 109, 110, and 111) yields the predictive incremental models

$$\bar{y}_{t+i} \approx \theta_i \Delta u_t + \Psi_i' s_t + v_i(t) \qquad (112)$$

$$\Delta u_{t+i-1} \approx \mu_{i-1}\Delta u_t + \Phi_{i-1}' s_t + v_i(t-1) \qquad (113)$$

Use of this model in the cost functional

$$J = \frac{1}{T} E \left\{ \sum_{i=0}^{T-1} [\bar{y}_{t+i+1}^2 + \rho(\Delta u_{t+i})^2] \right\} \tag{114}$$

yields the control law

$$\Delta u_t = - \frac{\sum_{i=1}^{T} \theta_i \Psi_i + \rho \sum_{1}^{T-1} \mu_i \Phi_i}{\sum_{i=1}^{T} \theta_i^2 + \rho(1 + \sum_{i=1}^{T-1} \mu_{i-1}^2)} s_t \tag{115}$$

Thus,

Algorithm 7 (MUSMAR incremental algorithm).

1. Using DFRLS estimate the parameters in the incremental models

$$\bar{y}_{i+i-T} = \theta_i \Delta u_{t-T} + \Psi_i' s_{t-T} + v_i^y(t-T) \tag{116}$$

$$\Delta u_{t+i-T-1} = \mu_{i-1} \Delta u_{t-T} + \Phi_{i-1}' s_{t-T} + v_i^u(t - T - 1) \tag{117}$$
$$i = 1, \ldots, T$$

2. Update the controller gains vector **F** by

$$\mathbf{F} = - \frac{\sum_{i=1}^{T} \theta_i \Psi_i + \rho \sum_{1}^{T-1} \mu_i \Phi_i}{\sum_{i=1}^{T} \theta_i^2 + \rho(1 + \sum_{i=1}^{T-1} \mu_{i-1}^2)} \tag{118}$$

and calculate the control by

$$u_t = u_{t-1} + \Delta u_t + \eta_t$$
$$\Delta u_t = \mathbf{F}' s_t \qquad\qquad \square$$

Admitting that the reference is constant, subtracting the reference from the output and expanding Δy_t on Equation (109) as $y_t - y_{t-1}$ results in the following pseudostate:

$$s_t = [\bar{y}_t \ldots \bar{y}_{t-n} \Delta u_{t-1} \ldots \Delta u_{t-m}] \tag{119}$$

If $n = 3$ and $m = 0$ an adaptive controller with PID structure is obtained.

Algorithm 8 (MUSMAR with PID structure).

• Execute the steps as in *Algorithm 7*, but using the pseudostate given by (119). $\qquad\qquad \square$

2.5.   The MIMO Case

Up to now, all the algorithms have been formulated for the SISO case. In order to cover the MIMO case, consider the unknown MIMO ARMAX plant

$$A(q^{-1})y(t) = B(q^{-1})u(t) + C(q^{-1})e(t) \tag{120}$$

where $u(t)$, $y(t)$ and $e(t)$ are, respectively, the input, output, and innovation vectors. $A(q^{-1})$, $B(q^{-1})$, and $C(q^{-1})$ are polynomial matrices in the unit delay operator q^{-1}, such that $B(0)$ is the null matrix A and B have no unstable common factors, C is strictly minimum-phase, i.e., the zeroes of each term of $C(z)$ are outside the closed unit circle and A, B, and C do not share common factors.

Associated with the plant a *quadratic cost functional* defined over a T-steps horizon is considered

$$E[J_T(t)] = \frac{1}{T} E[\|\tilde{y}_{t+1}^{t+T}\|_{Q_y}^2 + \|u_t^{t+T-1}\|_{Q_u}^2 | I^t] \tag{121}$$

where $T = 1, 2, \ldots$ and E denotes expectation conditioned on the information available at time t, I^t, w_t is an output reference vector to be tracked and

$$\gamma_{k-s}^k = [\gamma'(k)\ \gamma'(k-1)\ \ldots\ \gamma'(k-s)]'$$

$Q_y = Q_y' \geq 0$, $Q_u = Q_u' \geq 0$, $\|\gamma\|_Q^2 = \gamma'Q\gamma$. Q_y and Q_u are, respectively, output and control weight matrices.

The admissible control strategy amounts to selecting the inputs u_t according to a linear nonantecipative feedback law of the form $u_t = F's_t$ where s_t is the control information pattern or pseudostate:

$$s_t \triangleq [y_t' \ \ldots\ y_{t-ny+1}'\ u_{t-1}'\ \ldots\ u_{t-nu}'\ w_{t+T-1}']' \tag{122}$$

and assuming a constant feedback applied over the control horizon from time $t + 1$ up to $t + T - 1$. For unknown plants, this strategy is adaptively implemented by the following:

Algorithm 9 (positional MIMO MUSMAR).

1. Given all i/o data up to time t, recursively update the RLS estimate of θ_i, Ψ_i, μ_i, and Φ_i in the multipredictor model

$$\tilde{y}_{t+i-T} = \theta_i u_{t-T} + \Psi_i' s_{t-T} + v_i(t-T) \tag{123}$$

$$u_{t+i-T-1} = \mu_{i-1} u_{t-T} + \Phi_{i-1}' s_{t-T} + v_{i-1}(t-T) \tag{124}$$
$$i = 1, \ldots, T$$

$$v_i(t), \ v_i(t) \perp [u_t, \ s_t]$$

$$\bar{y}_{t+i-T} \triangleq y_{t+i-T} - w_{t+i-T}$$

2. Update the controller parameters by

$$\mathbf{F} = -\left[\sum_{i=1}^{T} (\theta_i' Q_y \theta_i + \mu_{i-1}' Q_u \mu_{i-1}) \right]^{-1}$$

$$\left[\sum_{i=1}^{T} (\theta_i' Q_y \Psi_i + \mu_{i-1}' Q_u \Phi_{i-1}) \right] \qquad (125)$$

and calculate the control by

$$u_t = F' s_t + \eta_t \qquad (126)$$

where $\{\eta_t\}$ is a vector white-noise dither signal of small variance, independent of $\{e_t\}$.

2.6. Constraining the Input

In practice, due to the actuator saturation imposed by the physical limits, the level of the input tension must be constrained. In order to minimize the conse-quences of these constraints, two actions are taken.

The first consists of feeding the regressor vectors not with u_t as calculated by $u_t = F' s_t + \eta_t$, but by $u_t = sat(F' s_t + \eta_t)$, in which "*sat*" is a convenient saturation fuction. In this way, the parameter estimator is fed with the correct plant input.

The second corrective action consists in suitably modifying the basic optimi-zation procedure, in order to contemplate the constraints. One simple possibility is to follow the line of [47], in which the cost to minimize is

$$J = E[y^2(t)] \qquad (127)$$

subject to the inequality constraint

$$E[u^2(t)] \leq c^2 \qquad (128)$$

Algorithm 10—Constrained input mean square-level MUSMAR.

1. At each sampling interval, recursively execute the steps in Algorithm 5.
2. Update the input weight ρ_t according to the Robbins–Monro scheme

$$\rho_{t+1} = \rho_t + \alpha_{t+1}\rho_t \frac{1}{c^2}[u_t^2 - c^2] \tag{129}$$

where c^2 is the constraint in the mean square value of the input, and $\{\alpha_t\}$ forms a sequence of positive scalars, such that $0 < \alpha_t < 1$.   □

2.7. Detecting Abrupt Changes

In this section, the basic MUSMAR algorithm is coupled with a fault-detection mechanism, in order to speed up the adaptation rate in response to a sudden change in plant dynamics. In robotics this happens (e.g., in manipulators picking up loads or starting to drill).

A number of fault-detection algorithms exist [48], but not all are adequate to the use in Adaptive Control. Basically, faults (i.e., abrupt changes in plant dynamics) may be detected by taking some measure of either the prediction-error sequence or the parameter-estimates sequence. The second alternative is preferred here, since it leads in practice to more sensitive and easily tunable algorithms. Specifically, MUSMAR is coupled with the fault-detection algorithm described in [40]. The rationale behind this method is the following: In RLS, at convergence, there is no correlation between the increments in successive instants of the parameter estimates since, otherwise the estimates could be improved. When there is some correlation, this means that the estimates are drifting in order to accommodate to the new value of the parameters, and, thus, that a fault has occurred. In order to explore this idea, the following quantities are built [40]:

$$w(t - 1) = \gamma_1 w(t - 2) + \Delta\hat{\theta}(t - 1) \tag{130}$$

$$\sigma(t) = sign[\Delta\hat{\theta}^1(t)w(t - 1)] \tag{131}$$

$$r(t) = \gamma_2 r(t - 1) + (1 - \gamma_2)\sigma(t) \tag{132}$$

where γ_1 and γ_2 are gains adjustable between zero and one, which define the rate of false alarms to misses. The quantity $w(t)$ is an exponential filtering of the increments of the estimates $\Delta\hat{\theta}(t)$ and, when a fault occurs, gives an estimate of the direction of the parameter change towards the new value. When the sequence $\sigma(t)$ is $+1$ frequently, it is concluded that a fault has occurred. In turn, this means that $r(t)$ will grow and cross some threshold. When this threshold is crossed a fault is detected.

The above procedure can be applied to any of MUSMAR predictors. The one-step ahead predictor for the output is, however, preferred, since its parameters are the only ones that depend on the open-loop plant. When a fault is detected, the diagonal of the covariance matrix is increased.

2.8.   Start-Up Procedures

Start-up should take place in the smoothest possible way. In order to avoid initial high-control actions, θ_1 (in MUSMAR) should be initialized to one. The regressors should contain plant data. The initial value of the covariance matrix is immaterial as long as it is high. Usually, $P(-1) = 1000I$ is a good choice. The reference should not present abrupt changes. Whenever possible, it should gradually be increased starting from the initial equilibrium point.

The use of Algorithm 9 can also contribute to a smooth start-up. The control weight tends usually to grow at the beginning, before converging to its final value. This initial growth reduces the feedback gains and helps reduce the start-up transients.

In some plants, where the noise is highly correlated (e.g., in drilling manipulators), there may be an initial strong transient that forces the entries of the covariance matrix to decrease very much. Consequently, the parameter estimates become frozen at values that may differ substantially from the optimal ones, which yields a very high cost. This phenomenon may be avoided by either doing a reset of the covariance matrix or by saturating the input during the initial period. A *normalization* implemented in a convenient way may also be useful.

The above discussion concerns the avoidance of start-up transients with *no* "a priori" knowledge about the plant being controlled. Often, this is not the case and an (at least stabilizing) estimate of the optimal feedback is available. The controller may, thus, be initialized with this feedback for a number of steps before using the feedback actually calculated by the adaptive algorithm.

2.9.   Stability and Convergence

As already referred to in the introduction, stability and convergence properties are crucial for adaptive algorithms. The nonlinear character of the equations defining the algorithms considered, and the difficulty in defining nominal solutions, prevents the use of most techniques [49]. Despite these difficulties, two kinds of results are available.

For IMRHAC with DFRLS, in the special case where the first T samples of the plant impulse responses are known and the system is embedded in a deterministic environment, a global stability and convergence result is available [36, 50]. The plant is supposed to be linear with a known upper-bound on its order. The proof technique is based on the *Key Technical Lemma* [51] and critically relies on a linear error equation.

For the analysis in the presence of unmodeled dynamics, the main theoretic result asserts that the MUSMAR algorithm, *if* it converges and the control horizon is large enough, converges close to the local minima of the quadratic cost functional [36, 52]. Simulations with reduced complexity controllers confirm the

theory and show the good performance of MUSMAR, even in difficult situations, specially designed as benchmark problems, for which other algorithms like
GPC fail to converge or converge with detuning. Here, the *model redundancy*
provided by the *separate* estimation of the predictors plays a major role. Indeed,
when unmodeled dynamics are present, the underlying ARMAX model is just an
approximation, and the extrapolation of a single-predictive model yields wrong
results. The separate estimation of the predictors, instead, is able to correctly
describe the behavior of the plant over the control horizon, and allows the good
performance of MUSMAR.

3.   ROBOT ARM MOTION CONTROL

This section is concerned with the application of the algorithms described in
Section 2 (the control of the motion of robot manipulators). Only MUSMAR is
considered, since it is believed to be the best of the controllers described. Application of IMRHAC to robot maniplators can be seen in [26], and the control of a
one-joint flexible arm is shown in [22].

The first example is concerned with a flexible arm with a reduced complexity
controller, and illustrates the robust convergence properties of MUSMAR.

The second and third examples are concerned with simulations on a two-link
rigid arm using, respectively, a MIMO and a SISO controller. The aim is to show
the ability MUSMAR has in coping with the nonlinearities of robotic manipulators, and to illustrate several features.

Finally, the fourth example shows tests conducted on a real manipulator.

3.1.   A Flexible Arm with a Reduced Complexity Controller

Experiment 1.
The following nonminimum phase, stable fourth-order ARMAX plant is an
approximate model of a flexible robot arm [2], which keeps the first two modes
(Figure 6.4):

$$y_{t+4} - 0.167y_{t+3} - 0.74y_{t+2} - 0.132y_{t+1} + 0.87y_t = \qquad (133)$$
$$= 0.132u_{t+3} + 0.545u_{t+2} + 1.117u_{t+1} + 0.262u_t + e_{t+4}.$$

Consider the restricted complexity controller defined by

$$u_t = f_1 y_t + f_2 y_{t-1}$$
$$\rho = 10^{-4}.$$

The cost as a function of the feedback gains results in a narrow valley and,
thus, constitutes a difficult minimization problem. Another classical difficulty,

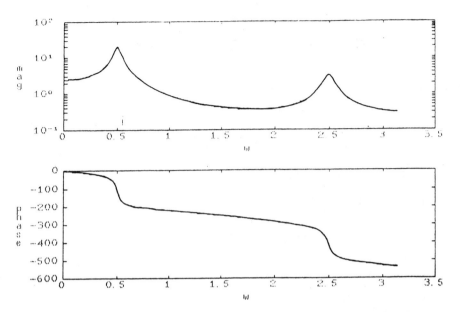

FIGURE 6.4. (Experiment 1) A Frequency Response of the Flexible Arm

which is present in robust adaptive control [40], comes from the fact that the frequency response of the plant presents a high-frequency resonant mode (see Figure 6.4).

Figures 6.5 and 6.6 show the convergence of the feedback calculated by MUSMAR, with $T = 5$, starting from open-loop. The final equilibrium point is quite close to the minimum. *All of the other controllers considered are unable to yield a stable closed-loop.*

3.2. Two-Link Rigid Arm—Simulations with SISO Controller

Consider a manipulator modeled as a set of n rigid links, connected in series, with friction acting at the joints. The dynamic equations for this system are developed by a direct application of classical mechanics [53], resulting in the following vector equation

$$\tau = M(\theta)\ddot{\theta} + V(\theta, \dot{\theta}) + G(\theta) \qquad (134)$$

Here, τ is the $n \times 1$ vector of joint moments and forces applied to the manipulator. $M(\theta)$ is a $n \times n$ symmetric matrix, relating the joint positions to the links inertia characteristics, called the manipulator mass matrix. $V(\theta, \dot{\theta})$ is a $n \times 1$ vector representing the forces and torques arising from centrifugal and friction effects. $G(\theta)$ is a $n \times 1$ vector, representing torques resulting from gravity terms.

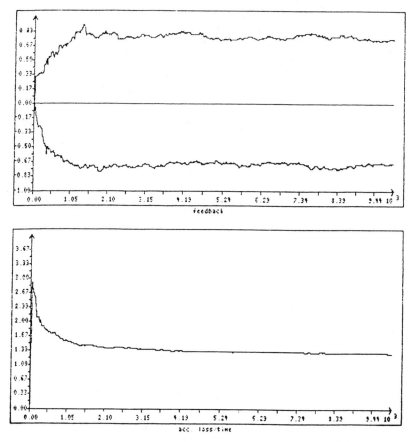

FIGURE 6.5.   (Experiment 1) The Time Evolution of the Feedback (above) and the Accumulated Loss Divided by Time (below), when the Flexible Arm is Controlled by MUSMAR with $T = 5$

A model of a two-link planar manipulator is used on the simulations (Figure 6.7). It is considered to have point masses (m_1 and m_2) at the end of each link (of lengths l_1 and l_2, respectively) and it includes gravity and friction terms.

In this particular case, Equation (134) results in the set of two nonlinear equations:

$$\tau_1 = m_2 l_2^2(\ddot{\theta}_1 + \ddot{\theta}_2) + m_2 l_1 l_2 c_2(2\ddot{\theta}_1 + \ddot{\theta}_2) + (m_1 + m_2)l_1^2 \ddot{\theta}_1 - m_2 l_1 l_2 s_2 \dot{\theta}_2^2$$
$$- 2m_2 l_1 l_2 s_2 \dot{\theta}_1 \dot{\theta}_2 + m_2 l_2 g c_{12} + (m_1 + m_2)l_1 g c_1 + v_1 \dot{\theta}_1 + k_1 sign(\dot{\theta}_1) \quad (135)$$

$$\tau_2 = m_2 l_1 l_2 c_2 \ddot{\theta}_1 + m_2 l_1 l_2 s_2 \dot{\theta}_1^2 + m_2 l_2 g c_{12} + m_2 l_2^2(\ddot{\theta}_1 + \ddot{\theta}_2)$$
$$+ v_2 \dot{\theta}_2 + k_2 sign(\dot{\theta}_2) \quad (136)$$

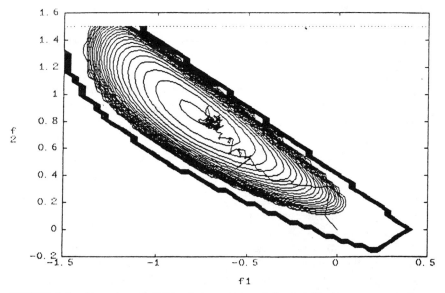

FIGURE 6.6. (Experiment 1) The Convergence of the Feedback Superimposed on the Level Curves of the Cost, when the Flexible Arm is Controlled by MUSMAR with $T = 5$

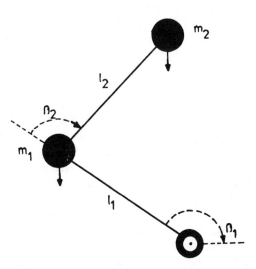

FIGURE 6.7. Manipulator Structure

where g is gravity acceleration, v_1 and v_2 are viscous friction coefficients, and k_1 and k_2 are Coulomb friction coefficients.

In the examples to be presented, $g = 9.8ms^{-2}$, $m_1 = m_2 = 0.5Kg$, $l_1 = l_2 = 0.5m$, $v_1 = v_2 = 1$, and $k_1 = k_2 = 0$.

Experiment 2.

Here the control structure used with the two-link manipulator has two MUSMAR, based SISO controllers, one acting at each link. The control horizon used is $T = 3$ and the regressor is of the form

$$s_t^i = [y_t^i \, y_{t-1}^i \, u_{t-1}^i w_{t+T}^i]' \tag{137}$$
$$i = 1, 2$$

where y_t^i is the joint position, u_t^i is the applied torque, and w_t^i is the reference for the joint position. The superscript i indicates to which joint the variables are referred.

This experiment shows a simulation of a manipulator acting on the horizontal plane (i.e., the gravity terms on the manipulator model are turned to zero). Figure 6.8 shows the adaptation transient, followed by a learning stage (up to 50 *sec*) and the manipulator movement to a predefined position. Both joint positions and their references are shown.

During the learning stage, several step changes of small amplitude are applied to the reference of Joint 1, in order to improve identification. Experience shows that when the algorithm has no "a priori" information available, starting both controllers at the same time is a very difficult task, due to the existing coupling. To avoid this problem, the controller on Joint 2 is started only sometime after the controller of Joint 1. Figure 6.8 shows that after the learning procedure is completed, the system has good tracking characteristics.

In these simulations several design parameters proved to be important:

- the forgetting factor (λ),
- the weight on the control cost (ρ), and
- the sampling rate.

As the dynamic characteristics of the system change with joint positions, the use of a forgetting factor is essential. The value of the forgetting factor depends on the system dynamics and the sampling rate used. The experiments presented are done using a constant forgetting factor. Values of $\lambda = 0.97$ for Joint 1 and $\lambda = 0.995$ for Joint 2 are used.

At the beginning, the information the controller has about the system dynamics is small, therefore, it is important to use some control weight in order to reduce oscillations during initial adaptation. In the simulations presented, ρ is made equal to 0.001 for the controller of Joint 1 and is equal to 0.0001 for the controller of Joint 2.

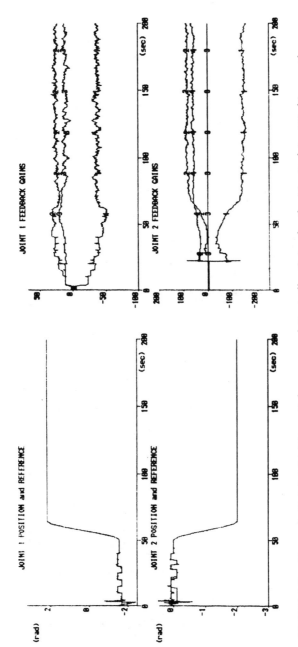

FIGURE 6.8. (Example 2) Joint Positions, References, and Controller Gains for the Manipulator Working on the Horizontal Plane. The Controller is Based on SISO MUSMAR Positional Algorithm

167

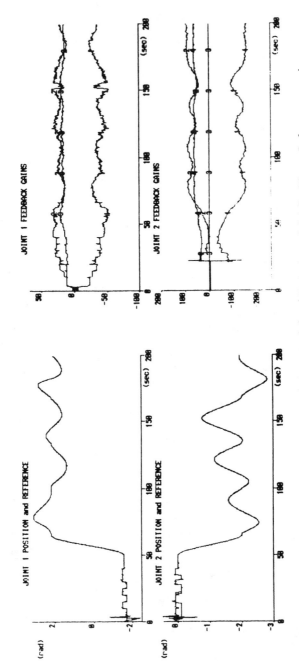

FIGURE 6.9. (Example 3) The Manipulator Describing a Predefined Trajectory: Joint Positions, References, and Controller Gains

168

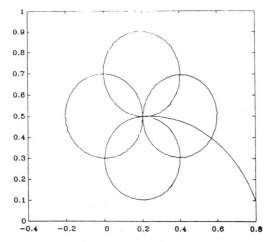

FIGURE 6.10. (Experiment 3) Trajectory of the Manipulators End Point (cartesian coordinates)

The sampling rate and the control horizon must be related. For the two link manipulator, with both controllers having a control horizon of $T = 3$, this leads to a sampling time of about 0.1 *sec* for the first joint and 0.025 *sec* for the second. Different sampling rates are required at the two joints because, when both controllers are active, the second joint must be sampled at a faster rate than the first one, since it "sees" a smaller inertia.

Example 3.

Figure 6.9 shows an experiment similar to the previous one, but here the arm describes the motion presented at Figure 6.10. As expected from the system dynamics, the gains change with the manipulator position.

Experiment 4.

The simulations, previously presented, refer to the case of a manipulator acting on the horizontal plane. If the manipulator is working on a vertical plane, the controller performance is worse than in the previous case, due to the effect of gravity and the integral effect that must be provided.

Hereafter *Algorithm 8* is used with a control horizon of $T = 3$, and a pseudo-state with $n = 2$, and $m = 0$.

Experiment 5.

In this simulation, two controllers of the type defined by Algorithm 8, act at the joints. The manipulator is working on the vertical plane. Figure 6.11 shows both joint positions and their references and Figure 6.12 the torques applied at the joints. During start-up the controller of the first joint is the only one active. Controller two starts during the training stage of Controller one. This can be noticed by the change of feedback-gain values. When the manipulator moves to a fixed position, the controller gains converge, after an adaptation transient.

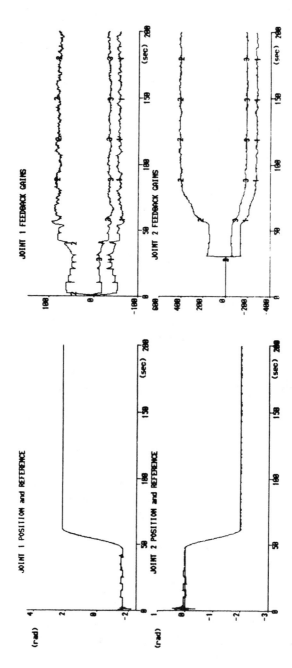

FIGURE 6.11. (Experiment 4) A Manipulator Working on the Vertical Plane. The Controller is Based on MUSMAR Incremental Algorithm: Joint Positions, References, and Controller Gains

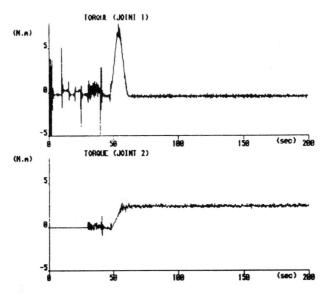

FIGURE 6.12. (Experiment 4) A Manipulator Working on the Vertical Plane: Torque Applied at the Joints

The performance improvement achieved with the controller based on the incremental algorithm, with respect to the one using the standard ("positional") algorithm, is considerable. The manipulator tracks the reference closely and there is no significant position static error.

Experiment 5.

Figure 6.13 presents a simulation similar to the previous one, but in which the manipulator executes the trajectory of Figure 6.15. Figure 6.14 shows the torques applied at the joints.

Experiment 6.

Figure 6.16 presents the case where, after adaptation, the manipulator picks a load, which represents a change in the end-link mass value. As shown in this figure, the joint error resulting from load change is very small and is quickly recovered, attesting the good performance of the controller. Observing the feedback gains for this simulation it is noticed that they change when the manipulator picks up the load.

The torques applied at the joints are shown in Figure 6.17, where the instant in which the manipulator picks up the load is easily identified by the change on the applied torques.

Experiment 7.

A situation, often found in industrial environments, is the case where the manipulator is submitted to disturbances, which are caused by the tools the manipulator is handling. This disturbance is modeled as band-pass noise applied

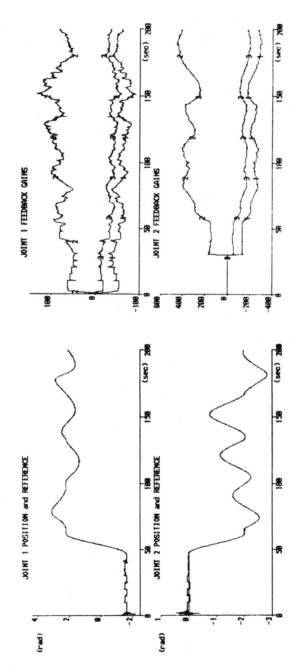

FIGURE 6.13. (Experiment 5) A Manipulator Describing a Predefined Trajectory on the Vertical Plane: Joint Positions, References, and Controller Gains

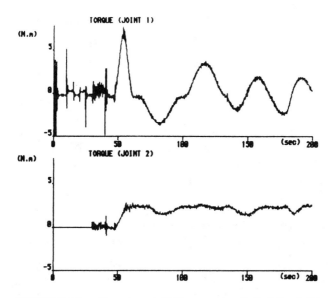

FIGURE 6.14.   (Experiment 5) Torques Applied at the Joints

at the terminal link, where the frequencies associated with the disturbance depend on the tool the manipulator is holding.

Figures 6.18 and 6.19 refer to the case where the disturbance applied is in the range from $0.8Hz$ to $1.7Hz$. In this situations, both controllers' feedback gains are different than those of Experiment 4. The torques applied by each controller

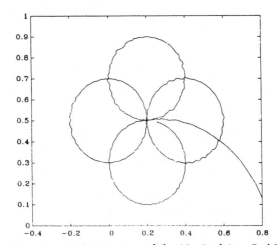

FIGURE 6.15.   (Experiment 5) Trajectory of the Manipulators End Point (cartesian coordinates) using MUSMAR with Integral Effect

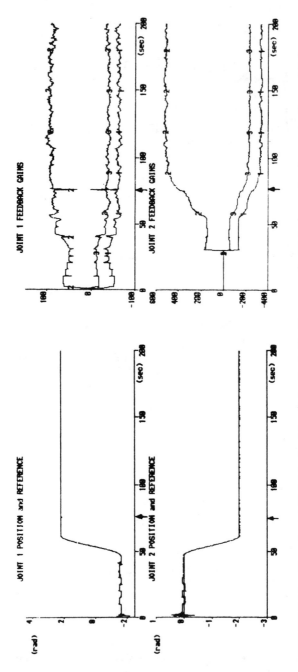

FIGURE 6.16. (Experiment 6) A Manipulator Behavior in the Presence of a Disturbance (a change on the end link mass): Joint Positions, References, and Controller Gains. The Manipulator Picks up a Load at a Time = 75sec (marked by the arrow). Notice the Change in the Gain Values when the Load is Picked up

174

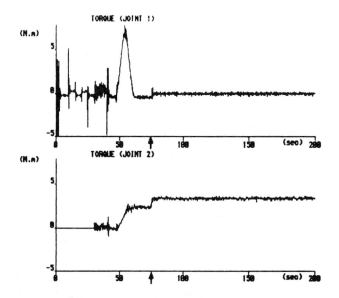

FIGURE 6.17. (Experiment 6) Torques Applied at the Joints. The Torques Increase when the Manipulator Picks up the Load

are also different from the ones of Figure 6.13. However, the manipulator shows no loss of performance.

Experiment 8.

One of the problems in using adaptive controllers is the need of a learning stage which results from the fact that when the controller starts it has no information on the system dynamics. One way to avoid this problem is starting the controller with information obtained from previous experiments.

Figures 6.20 and 6.21 show the result of a simulation where controller feedback gains are initialized using values obtained in previous experiments, and are kept constant during the first 40 *sec*. After that period the feedback gains are calculated using the adaptive algorithm.

Figure 6.22 shows a simulation where the feedback gains are kept constant through the entire simulation. It clearly shows that adaptation is essential in order to stabilize the manipulator.

3.3. Two-Link Rigid Arm—Simulations with MIMO Controller

This section reports simulation experiments performed on the multivariable long-range adaptive control of a rigid two-link robot arm.

The control structure used with the two-link manipulator is a MUSMAR

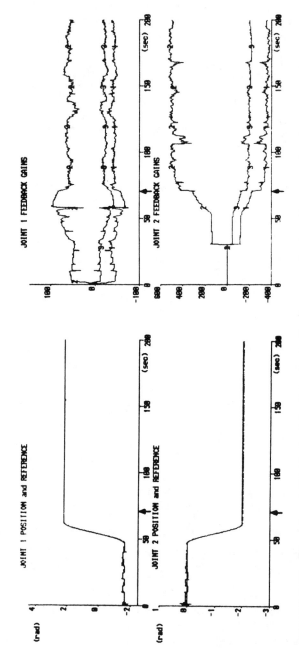

FIGURE 6.18. (Experiment 19) Manipulator Behavior in the Presence of a Low Frequency Band-Pass Disturbance: Joint Positions, References, and Controller Gains. The Disturbance Starts at Time = 70sec

176

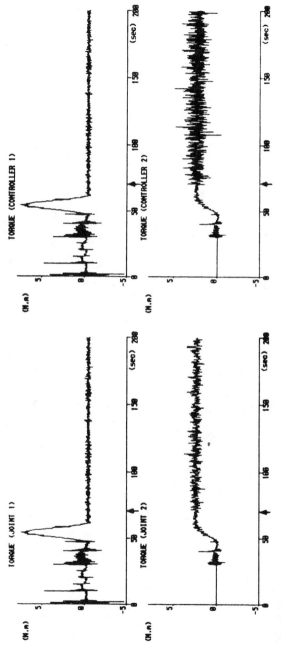

FIGURE 6.19. (Experiment 19) (A) Torques Applied at the Joints (equal to the sum of the torques computed by the controller and the disturbance torques) for the Case where a Low Frequency Band-Pass Disturbance is Present. (B) Torques Computed by the Controller

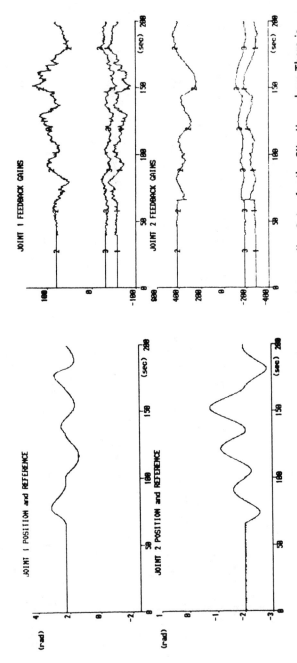

FIGURE 6.20. (Experiment 8) Joint Positions, their References, and Controller Gains for the Situation where There is a Smooth Start-Up, Resulting from Initialising the Controller with Information from Previous Experiments. Note that the Gains are Kept Constant During the Initial 40 Seconds

178

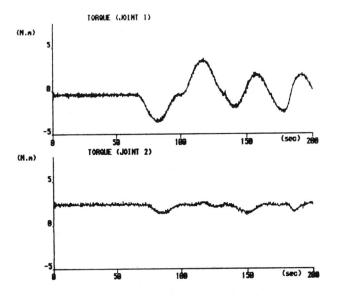

FIGURE 6.21.   (Experiment 8) Torques Applied at the Joints

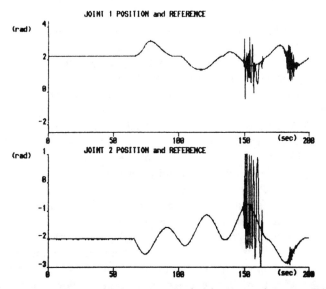

FIGURE 6.22.   (Experiment 8) Joint Positions for the Case of Constant Feedback Gains. In this Experiment the Gains are Kept Constant. Their Values are the Same Used to Initialize the Controller for Smooth Start-Up

multivariable controller (Algorithm 9) with two inputs and two outputs. The control horizon used is $T = 3$ and the pseudostate is of the form

$$s_t = [\bar{y}_t' \; \bar{y}_{t-1}' \; \bar{y}_{t-2}' \; w_{t+T}']'$$
$$\bar{y}_i \triangleq y_i - w_i \tag{138}$$

where y_t is the joint's position vector, u_t is the applied torque vector, and w_t is the reference vector for the joint's position.

The output and control weight matrices are of the form

$$Q_y = \begin{bmatrix} 1 & 0 \\ 0 & 1 \end{bmatrix} \tag{139}$$

$$Q_u = \begin{bmatrix} \rho_1 & 0 \\ 0 & \rho_2 \end{bmatrix} \tag{140}$$

where ρ_1 and ρ_2 are exponentially decreasing values ranging from .0005 to .00001 and from .0005 to .0002, respectively. The variation on the control weights is performed during a 20 *sec.* learning stage, when the references consist of low-amplitude PRBSs for identification. Better results are obtained when the PRBSs have slightly different periods (0.81 *sec.* and 0.79 *sec.* were the values used, respectively, for Joint 1 and Joint 2 reference PRBS periods). In order to allow a constant information renewal for identification, a small-variance dither was added to the control signals.

The LS estimates are computed using directional forgetting with the information matrix propagated by a UD factorization algorithm.

In order to achieve the results described next, special care is needed when tuning some parameters of the controller. The main difficulties of the algorithm consist of identifying the system, which is strongly nonlinear and has linearized parameters that change very quickly. For this purpose, it is necessary to use the *directional forgetting* version of RLS, as well as initial learning reference consisting of a PRBS signal. Since the robot arm can easily reach some nearly unstable equilibrium situations (for example, when the manipulator is pointing up, vertically), control problems may arise when the parameters of the identified model are not correct. A critical situation arises when the control is turned on in the learning stage and there is still no knowledge about the system. For this particular case, an exponentially decreasing control weight is used in the algorithm, varying from a high value in the beginning, in order to prevent a blow-up of the output of the controlled system, to a smaller value, after the learning stage, allowing a better reference tracking. After the learning stage, each link reference is composed of a third degree polynomial section followed by a sinusoidal function.

As referred to before, a non-zero control weight matrix is needed to avoid the controller turning unstable. Although the behavior of the horizontal plane manipulator is not affected by the application of the control weight (the output can easily follow the reference), the same does not happen when the robot arm works in the vertical plane. Here, a significant tracking error arises, because a positional algorithm is being used. Tests were made with the algorithm modified in order to use incremental values of the inputs instead of the inputs themselves, but this leads to an increase of the unstability of the whole system, and the results were not satisfactory.

The figures shown in the following experiments have the following notation:

- y^i—Position of Joint i,
- w^i—Reference of Joint i,
- $\bar{w}^i$—Virtual Reference of Joint i,
- ψ_{nm}—Element of Ψ_1 relating the position of joint m at time $t-1$ with position of Joint n at time T, and
- F_{nm}—Controller Gain Relating the Torque to be Applied at Joint n to the mth Element of s_t.

Experiment 9.

When the manipulator is in the horizontal position, the effect of gravity is null and the only variation on the parameters of the robot arm is due to the inertia of the manipulator that depends on the position of Joint 2. The response of the robot arm to a reference change is seen on Figure 6.23. During the first 20 seconds, which correspond to the learning phase for identification, there is a strong oscillation due to the almost integral character of the manipulator. Then, it can be noticed that the positional error of both joints becomes close to zero, what at first sight could mean that the parameters of the closed-loop system are correctly identified. Nevertheless, Figure 6.24, shows that ψ_{21} is kept at a low value, meaning that the effect of the position of Joint 1 in the behavior of Joint 2 is small (which is to be expected in the horizontal plane manipulator), and that the other parameters (ψ_{11}, ψ_{22}, ψ_{21}) are slightly sliding away from a constant value. Looking at the controller gains (Figures 6.25 and 6.26), it seen that their values are not stabilized, meaning that there is some redundancy in this set of gains. One reason for this to happen may be suggested by the fact that one of the values of the **D** matrix quickly goes to zero, while some other terms are kept within high values. This denotes a severe identification problem, which has already been minimized by using a directional forgetting factor.

3.4. Two-Link Rigid Arm—Tests with a Real Plant

This section is concerned with tests of the MUSMAR algorithm performed on a *real* two-link rigid arm. The arm used is a laboratory scale ARMATROL model ESA 101 (Figure 6.27) with four joints, one in the basis along the vertical axis and the other three in the same vertical plane. The joints are moved by DC mo-

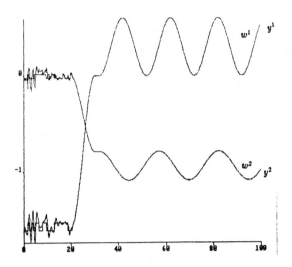

FIGURE 6.23. (Experiment 9) Joint Positions and References

tors and reduction gearboxes with a ratio of 1/600. For the tests, Joints 1 and 2 were selected, such as to form a planar arm working on the vertical plane.

Although other versions have been tested with similar results, only the tests with the positional version of MUSMAR are reported here.

Each joint is controlled independently, with a replica of MUSMAR. The start-up of the algorithm comprises the following stages: During the first ten iterations a constant electric tension of small amplitude is applied to the amplifiers of each

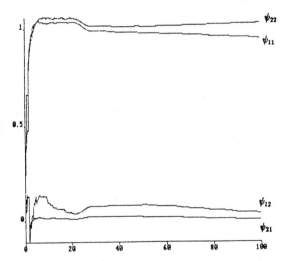

FIGURE 6.24. (Experiment 9) Some Identified Parameters of Ψ_1

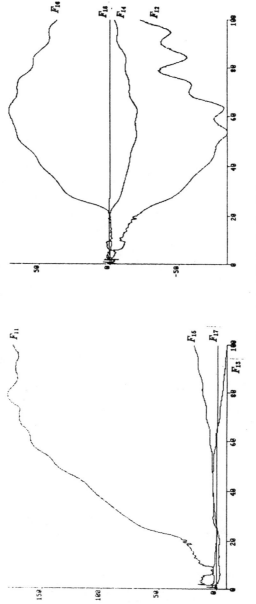

FIGURE 6.25. (Experiment 9) Controller Gains

183

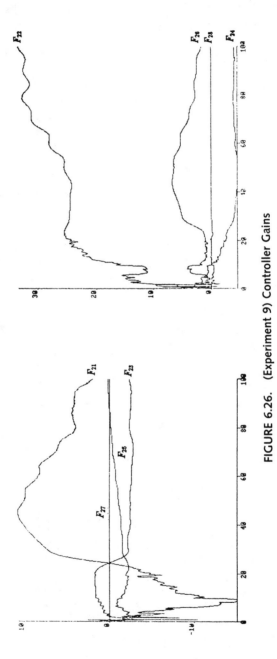

FIGURE 6.26. (Experiment 9) Controller Gains

184

FIGURE 6.27.   The Arm Used in the Examples of Section 3.4

joint. There is then a learning stage in which the reference to follow is a square wave of small amplitude. This procedure is followed first for the elbow joint and then for the shoulder joint. It is only when it is completed that normal operation starts.

Standard RLS with a variable forgetting factor are used to identify the predictive models. The forgetting factor is calculated according to [39]:

$$\lambda(t) = 1 - [1 - \varphi^T(t)K(t)]\epsilon^2(t)/\Sigma_0 \tag{141}$$
$$\lambda(t) < \lambda_{min} \Rightarrow \lambda(t) = \lambda_{min}$$

where $\varphi(t)$ is the regressor and $K(t)$ is the Kalman gain in the RLS equations, λ_{min} and Σ_0 are constants and ϵ is the prediction error (usually the one step-ahead predictor is used).

The reference trajectory of (1) is calculated with $\gamma = 0$.

Experiment 10.

This experiment uses $T = 3$,

$$s_t = [y_t \; y_{t-1} \; y_{t-2} \; u_{t-1} \; u_{t-2} \; w_{t+T}]'$$

$\gamma^1_{min} = \lambda^2_{min} = 0.996$, $\Sigma^1_0 = \Sigma^2_0 = 5000$, $\rho_1 = \rho_2 = 0.0001$. Figure 6.28 shows the results obtained. Note the evolution of the forgetting factor whose value raises, becoming close to one and "freezing" the information obtained, whenever the reference is constant. After 200 *seconds* there is an abrupt change in the gains

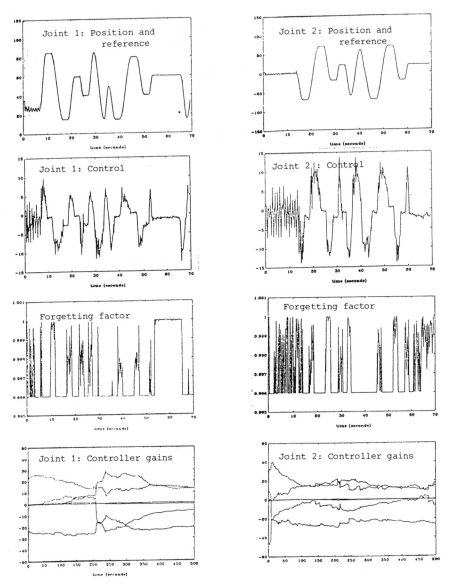

FIGURE 6.28.   Experiment 10

for Joint 1, which causes a loss in performance. As shown in the next example, this is avoided by increasing the horizon T.

Experiment 11.

This experiment illustrates the fact that an increase in the horizon T leads to a better convergence of the controller gains. The structure of the controller is equal

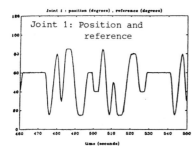

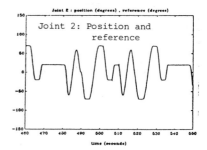

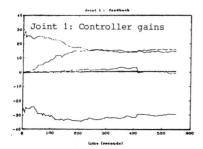

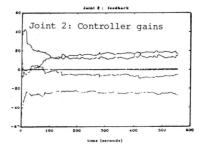

FIGURE 6.29.   Experiment 11

to the one of Experiment 10, except that T is made equal to five. Figure 6.29 shows the results obtained for both joints. Comparing with Figure 6.28, the improvement in gain convergence is neat.

With $T = 1$ (corresponding to the controller described in [16]), the controller presents frequent oscillations due to the drift of the gains, and can easily become unstable.

Experiment 12.

This experiment illustrates the response of the algorithm to sudden load changes. The configuration used is the same of Experiment 10.

Figure 6.30 shows the results obtained. During the first 50 *seconds* the reference changes, it is then kept constant. At $t = 240$ *seconds* the arm is loaded with a mass of $1/2$ *kg* (maximum allowed), which is removed after one minute.

4.   MOTION CONTROL OF AN UNDERWATER ROV

This section is concerned with the adaptive control of the motion of an underwater ROV (Remote-Operated Vehicle). The depth and pitch control subsystems

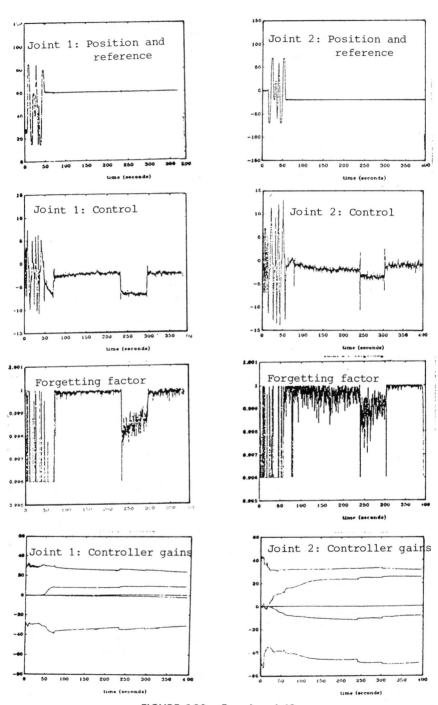

FIGURE 6.30.   Experiment 12

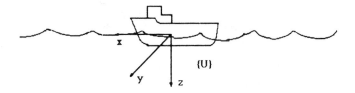

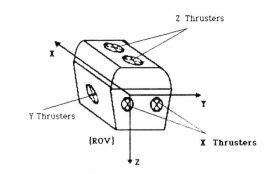

FIGURE 6.31.   The Coordinate Systems in the ROV Model

are considered. The examples presented are based on a model of the ROV dynamics (see [54] for details and further references).

The model considers two coordinate systems $\{U\}$ and $\{ROV\}$ attached, respectively, to the mother-ship and the ROV (Figure 6.31). It is assumed that $\{U\}$ is an inertial coordinate system. All the variables are expressed in $\{ROV\}$ coordinates and the notation of [55] is used.

In order to develop the model, the equations describing the motion of the rigid body have been used (third law of Newton and Euler equations [56]). The force and moment exerted by the water on the body are given using the concept of added masses [57, 58]. These forces correspond to the ones acting on a body immersed in a nonviscous fluid and their effect is to increase the inertial of the body. The model Figure 6.32 is expressed in the following equations:

$$\dot{U} = \mathcal{Y}^{-1}\{F - \mathcal{M}_{TR}\mathcal{M}_R^{-1}|N - \Omega \times (\mathcal{M}_R\Omega + \mathcal{M}_{RT}U) - U \times (\mathcal{M}_{TR}\Omega)$$
$$- A_{RT}(\Omega \times U_W) + \Omega \times (A_{RT}U_W) + U_W \times (A_{TR}\Omega)]$$
$$- \Omega \times (\mathcal{M}_TU + \mathcal{M}_{TR}\Omega) - A_T(\Omega \times U_W) + \Omega \times (A_TU_W)\}$$

$$\dot{\Omega} = \mathcal{X}^{-1}\{N - \mathcal{M}_{RT}\mathcal{M}_T^{-1}[F - \Omega \times (\mathcal{M}_TU + \mathcal{M}_{TR}\Omega) - A_T(\Omega \times U_W)$$
$$+ \Omega \times (A_TU_W)] - \Omega \times (\mathcal{M}_R\Omega + \mathcal{M}_{RT}U) - U \times (\mathcal{M}_{TR}\Omega)$$
$$- A_{RT}(\Omega \times U_W) + \Omega \times (A_{RT}U_W) + U_W \times (A_{TR}\Omega)\}$$

$$\overset{U}{_{ROV}}\dot{R} = \{(\overset{U}{_{ROV}}R\Omega) \times\} \overset{U}{_{ROV}}R$$

$$d/dt(^UP_{ROVORG}) = \overset{U}{_{ROV}}RU$$

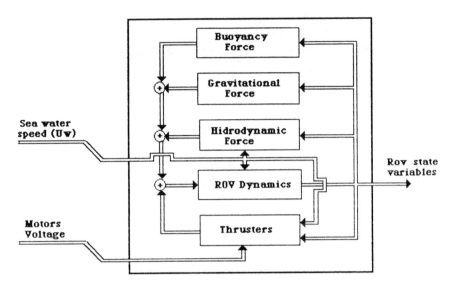

FIGURE 6.32.   Block Diagram of the ROV Model

where

- $\mathcal{Y}$, $\mathcal{Y}$ are matrices depending on the generalized mass matrix,
- F, N are the external forces and moments acting on the ROV,
- Ω is the angular velocity of the ROV expressed in $\{ROV\}$,
- U is the velocity of the ROV expressed in $\{ROV\}$,
- $^{U}_{ROV}R$ is the rotation matrix relating $\{ROV\}$ with $\{U\}$,
- U_W is the water velocity in $\{ROV\}$,
- $\mathcal{M}_R$, $\mathcal{M}_{RT}$, $\mathcal{M}_{TR}$, $\mathcal{M}_R$ are the generalized inertia matrices of the ROV, and
- A_T, A_{RT}, A_{TR}. A_R are the added masses matrices of the ROV, and
- $^{U}P_{ROVORG}$ is the position of the $\{ROV\}$ origin in $\{U\}$.

The external forces and moments acting on the ROV are

$$F = F_W + F_B + F_H + F_T$$
$$N = N_W + N_B + N_H + N_T$$

where

- F_W, N_W are the forces and moments due to the weight of the ROV,
- F_B, N_B are the forces and moments due to buoyancy,
- F_H, N_H are the hidrodynamic (drag) forces and moments [59] assuming the vehicle operating under turbulent flow, these forces are approximated by the

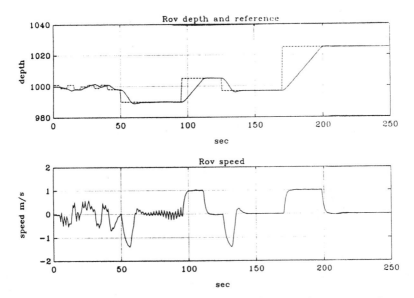

FIGURE 6.33. **(Experiment 13) ROV Depth and Reference when Positional MUSMAR is Used**

square-law resistance $F = Ku|u|$, in which u is the velocity of the ROV and K is a constant, and

- F_T, N_T are the forces and moments due to thrusters.

Experiment 13.

In this experiment, the depth of the ROV is controlled using Algorithm 5 (MUSMAR positonal controller) with $T = 3$, $\rho = 0.0001$, a sampling interval of 0.65 *seconds* and $na = 3$, $nb = 2$, $nw = 1$. Figure 6.33 shows the results obtained evidencing a good performance. Despite the fact that a positional algorithm is being used, the controller achieves an almost zero steady-state error.

Experiment 14.

In this experiment, the pitch of the ROV is controlled using Algorithm 8 (MUSMAR with integral effect) with $T = 6$, $\rho = 0.00001$, a sampling interval of 0.65 *seconds,* and $n = 5$, $m = 4$. Figure 6.34 shows the results obtained.

5. CONCLUSION

A tutorial on the application of long-range adaptive controllers to robot motion control has been presented. The basic theory leading to the algorithms is reviewed, together with implementational aspects. A number of application examples show the effectiveness of the techniques described.

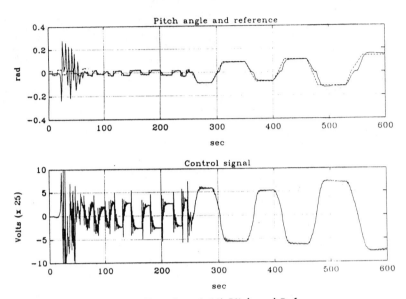

FIGURE 6.34. (Experiment 14) Pitch and Reference

REFERENCES

1. S. Dubowsky, and R. Kornbluh, "On the Development of High Performance Adaptive Control Algorithms for Robotic Manipulators," *Robotics Research*—The Second Internatonal Symposium," MIT Press, Cambridge, MA, 1985.

2. I.D. Landau, "Adaptive Control Techniques for Robotic Manipulators—The Status of the Art." *SYROCO-85, IFAC Symposium on Robot Control,* Barcelona, Spain, November 6–8, 1985.

3. J.J. Craig, *"Adaptive Control of Mechanical Manipulators."* Addison-Wesley, MA, 1988.

4. P. Lopez, and C. Maire, "Adaptive Methodologies and Robotics," *Proc. 25th IEEE CDC,* Athens, Greece, 1986.

5. S. Dubovsky, and D.T. DesForges, "The Application of Model Reference Adaptive Control to Robotic Manipulators." *Trans. ASME, J. Dynamic Systems, Measurement, and Control,* Volume 101, September 1979, pp. 193–200.

6. K. Lim and M. Eslami, "Robust Adaptive Controller Designs for Robot Manipulator Systems." *IEEE J. of Robotics and Automation* Vol. RA-3, No. 1, February 1987, pp. 54–66.

7. S.H. Murphy and G.N. Saridis, "Experimental Evaluation of Two Forms of Manipulator Adaptive Control." *Proc. 26th IEEE CDC,* Los Angeles, U.S.A., 1987, pp. 216–221.

8. M.B. Leahy, S.H. Murphy, R. Al-Jaar, and G.N. Saridis, "Experimental Evaluation of a PUMA Manipulator Model Referenced Adaptive Controller." *Proc. 26th IEEE CDC,* Los Angeles, U.S.A., 1981, pp. 2196–2201.

9.   H.R. Asare and D.G. Wilson, "Evaluation of Three Model Reference Adaptive Control Algorithms for Robotic Manipulators." *Proc. 1987 IEEE Int. Conf. on Robotics and Automation*, 1987, pp. 1531–1542.

10. I.H. Mufti, "Model Reference Adaptive Control for Manipulators—A Review." *Prep. SYROCO '85—IFAC Symp. on Robot Control*, Barcelona, Spain 1985, pp. 27–32.

11. P.K. Khosla and T. Kanade, "Experimental Evaluation of Nonlinear Feedback and Feedforward Control Schemes for Manipulators." *Int. J. Robotics Research*, Vol. 7, No. 1, 1988, pp. 18–28.

12. C.S.G. Lee, M.B. Chung, and B.H. Lee, "An Approach of Adaptive Control for Robot Manipulators." *J. Robotic Systems*, Vol. 1, No, 1, 1984, 27–57.

13. C.G. Atkeson and J. McIntyre, "An Application of Adaptive Feedforward Control to Robotics." *Prep. 2nd IFAC workshop on Adaptive Systems in Control and Signal Processing*, Lund, Sweden, 1986, pp. 137–142.

14. H. Seraji, "An Adaptive Cartesian Control Scheme for Manipulators." *Proc. 1987 IEEE Int. Conf. on Robotics and Automation*, 1987, pp. 157–164.

15. C.H. An, C.G. Atkeson, J.D. Griffiths, and J.M. Hollerbach, "Experimental Evaluation of Feedforward and Computed Torque Control." *Proc. 1987 IEEE Conf. on Robotics and Automation*, 1987, pp. 165–168.

16. A. Koive and T.-H. Guo, "Adaptive Linear Controller for Robotic Manipulators" *IEEE trans. Aut. Control*, Vol. AC-28, No. 2, 1983, pp. 162–170.

17. G.G. Leininger, "Adaptive Control of Manipulators using Self-Tuning Methods." *Robotics Research—The First International Symposium*, MIT Press, Cambridge, MA, 1984, pp. 801–816.

18. N. Honshaugi and A.J. Koivo, "Eigenvalue Assignement and Performance Index Based Force-Position Control with Self-Tuning for Robotic Manipulators." *Proc. 1987 IEEE int. conf. on Robotics and Automation*, 1987, pp. 1386–1391.

19. J.-J.E. Slotine and W. Li, "On the Adaptive Control of Robot Manipulators." *Int. J. Robotics Research*, Vol. 6, No. 3, 1987, pp. 49–59.

20. J.-J.E. Slotine and W. Li, "Adaptive Manipulator Control: A case study." *IEEE trans. Aut. Control*, Vol. 33, No. 11, 1988, pp. 995–1003.

21. C. Greco, G. Menga, E. Mosca, and G. Zappa, "Performance Improvements of Self-Tuning Controllers by Multistep Horizons: The MUSMAR approach." *Automatica*, Vol. 20, No. 20, 1984, pp. 681–699.

22. M. Lambert, *"Adaptive Control of Flexible Systems."* Ph.D. thesis, Oxford University, England, 1987.

23. M.A. Lelić and M.B. Zarrop, "Generalized Pole-Placement Self-Tuning Controller. Part 1. Basic Algorithm." *Int. J. Control*, Vol. 46, No. 2, 1987, pp. 547–568.

24. M.A. Lelić and P.E. Wellstead, "Generalized Pole-Placement Self-Tuning Controller. Part 2. Application to Robot Manipulator Control." *Int. J. Control*, Vol. 46, No. 2, 1987, pp. 569–601.

25. H.-B. Kuntze, A. Jacubasch, J. Richalet, and Ch. Arber, "On the Predictive Functional Control of an Elastic Industrial Robot." *Proc. 25th IEEE CDC*, Athens, Greece, 1986, pp. 1877–1881.

26. A. Soeterboek, H. Verbruggen, and P. Bosch, "An Application of a Self-Tuning Controller on a Robotic Manipulator." *Proc. 1987 IEEE int. conf. on Robotics and Automation*, 1987, pp. 1380–1385.

27. F.J. Coito and J.M. Lemos, "A Long-Range Adaptive Controller for Robot Manipulators." *Int. J. Robotics Research*, Vol. 10, No. 6, December 1991.

28. K. Yoshimoto and K. Wakatsuki, "Application of the Preview Tracking Control Algorithm to Servoing a Robot Manipulator." *Robotics Research—The Second International Symposium*, MIT Press, Cambridge, MA, pp. 883–897.

29. C.E. Rohrs, L. Valavani, M. Athans, and G. Stein, "Robustness of Continuous-Time Adaptive Control Algorithms in the Presence of Unmodeled Dynamics." *IEEE Trans. Aut. Contr.*, Vol. AC-30, No. 9, September 1985, pp. 881–889.

30. J.M. Lemos and E. Mosca, "A Multipredictor Based LQ Self-Tuning Controller." *Proc. 7th IFAC Symp. Ident. Syst. Param. Est.*, York, U.K., pp. 137–142.

31. E. Mosca and G. Zappa, "ARX Modeling of Controlled ARMAX Plants and its Application to Robust Multipredictor Adaptive Control." *Proc. 24th IEEE CDC*, Ft. Lauderdale, FL, U.S.A. 1985, pp. 856–861. Also, *IEEE Trans. Aut. Contr.*, January 1989.

32. D.W. Clarke, C. Mohtadi, and P.S. Tuffs, "Generalized Predictive Control—Part I. The Basic Algorithm." *Automatica*, Vol. 23, No. 2, 1987, pp. 137–148.

33. D.W. Clarke, C. Mohtadi, and P.S. Tuffs, "Generalized Predictive Control—Part II. Extensions and Interpretations." *Automatica*, Vol. 23, No. 2, 1987, pp. 149–160.

34. R.M.C. De Keyser, Ph. G.A. Van de Velde, and F.A.G. Dumortier, "A Comparative Study of Self-Adaptive Long-Range Predictive Control Methods." *Prep. 7th IFAC Symp. Ident. and Syst. Param. Est.*, 1985, pp. 1317–1322, York, U.K.

35. D.W. Clarke and C. Mohtadi, "Properties of Generalized Predictive Control." Technical Report, Oxford University, England, 1988.

36. J.M. Lemos, "*Adaptive LQ control.*" Ph.D. Thesis. Technical University of Lisbon, Portugal, 1989.

37. G. Casalino, F. Davoli, R. Minciardi, and G. Zappa, "On Implicit Modeling Theory: Basic Concepts and Application to Adaptive Control." *Automatica*, Vol. 23, No. 2, 1986, pp. 189–201.

38. T. Kailath, "*Linear Systems,*" Prentice-Hall, NJ, 1982.

39. T.R. Fortescue, L.J. Kershenbaum, and B.E. Ydstie, "Implementation of Self-Tuning Regulators with Variable Forgetting Factors." *Automatica*, Vol. 17, No. 6, 1981, pp. 831–835.

40. T. Hägglund, "*New Estimation Techniques for Adaptive Control,*" Ph.D. Thesis, Lund University, Lund, Sweden, 1983.

41. R. Kulhavý and M. Kárny, "Tracking of Slowly Varying Parameters by Directional Forgetting." *Proc. 9th IFAC World Congress*, Budapest, Hungary, 1984, pp. 79–83.

42. R. Kulhavy, "Restricted Exponential Forgetting in Real-Time Identification." *Automatica*, Vol. 23, No. 5, 1987, pp. 589–600.

43. G.J. Bierman, "*Factorization Methods for Discrete Sequential Estimation*, Academic Press, NY, 1977.

44. L. Chisci and E. Mosca, "Parallel Architectures for RLS with Directional Forgetting." *Int. J. Adaptive Control and Signal Processing*, Vol. 1, No. 1, 1987, pp. 69–88.

45. K.J. Åstrom, "Theory and Applications of Adaptive Control—A Survey." *Automatica* Vol. 19, 1983, pp. 471–487.

46. D.W. Clarke and P.J. Gawthrop, "Self-Tuning Controller." *Proc. IEE* Journal, Vol. D, No. 126, 1975, pp. 633–640.

47. H.T. Toivonen, "Variance Constrained Self-Tuning Control." *Automatica,* Vol. 19, No. 4, 1983, pp. 415–418.

48. M. Basseville, and A. Benveniste, *"Detection of Abrupt Changes in Signals and Dynamical Systems."* Springer-Verlag, NY, 1980.

49. Anderson, B., R. Bitmead, C. Johnson, P. Kokotovic, R. Kosut, I. Mareels, L. Praly, and B. Riedle, *"Stability of Adaptive Systems."* MIT Press, Cambridge, MA, 1986.

50. J.M. Lemos and E. Mosca, "Stability and Convergence of Multipredictor Adaptive Controllers." *Proc. 27th IEEE CDC* 1988.

51. G.C. Goodwin and K.S. Sin, *"Adaptive Filtering, Prediction and Control,"* Prentice-Hall, NJ, 1984.

52. E. Mosca, G. Zappa, and J.M. Lemos, "Robustness of Multipredictor Adaptive Regulators: MUSMAR." *Automatica,* Vol. 25, No. 4, 1989, pp. 521–529.

53. R.P. Paul, *"Robot Manipulators."* MIT Press, Cambridge, MA, 1981.

54. C. Silvestre, *"Modeling and Adaptive Control of an Underwater ROV."* Technical report. CAPS, Technical University of Lisbon, Portugal, 1989.

55. J.J. Craig, *Introduction to Robotics Mechanics and Control.* Addison-Wesley, Reading, MA, 1986.

56. K.R. Symon, *Mechanics.* Addison-Wesley, Amsterdam, 1971.

57. L. LandWeber, "Motion of Immersed and Floating Bodies," *Handbook of Fluid Dynamics*, McGraw Hill, NY, 1961.

58. H. Lamb, *Hydrodynamics.* Cambridge, NY, 1932.

59. L. Prandtl and O.G. Tietjens, *Applied Hydro- & Aeromechanics.* Dover, NY, 1957.

7

Research Problems in Computer Networking for Manufacturing Systems*

Asok Ray
Department of Mechanical Engineering
The Pennsylvania State University
University Park, PA

Shashi Phoha
Applied Research Laboratory
The Pennsylvania State University
University Park, PA

1. INTRODUCTION

The basic concepts and the goal of *Computer-Integrated Manufacturing (CIM)* have been introduced in [1]. The chapter also points out the need for an integrated communications network over which the information leading to operations of a total manufacturing system will flow. The major motivation for computer networking, as it is applied to manufacturing, is to significantly improve productivity and quality by eliminating the inefficiency of manual actions. The architecture and performance of such a network are of paramount importance to the efficient and reliable operation of CIM processes. Lack of interoperability, reliability, and flexibility in the network design can cause chain reactions of delay and congestion resulting in loss of control of safety and productivity.

Although the technology of computer networking is quite advanced, its specific application to CIM has not yet been well characterized. To this effect the Manufacturing Automation Protocol (MAP) [2] and the Technical and Office Protocols (TOP) [3] are being considered as standards for networking of integrated manufacturing and information processing systems by a number of leading domestic and foreign industries. However, there are many unresolved issues, such as interoperability, performance in real-time environment, and fault toler-

---

* The authors thankfully acknowledge the contributions of Dr. S. Aggarwal, Dr. K.H. Murali, Dr. R. Denton, Dr. A. Ephremides, and Dr. K.G. Shin as well as those of other participants of the workshop on *Computer Networking for Manufacturing Systems* [4].

ance and error recovery, related to MAP and TOP. Examples of other problems are resource scheduling, flow and congestion control, and operational stability of communication networks within the CIM environment. These problems must be addressed in advance by systematic analytical and experimental research.

To meet this goal the division of design, manufacturing and computer engineering of the National Science Foundation sponsored the workshop on *Computer Networking for Manufacturing Systems* [4]. The objectives of the workshop were to assess the state-of-the-art, identify and characterize research needs, and to recommend directions for future research in CIM networking. The workshop brought together different views on computer networking with the objective of enhancing manufacturing productivity. This chapter describes pertinent research topics which were recommended by different specialist groups in the workshop.

The article is organized in three sections including the introduction. The second section identifies five broad areas and provides recommendations for research in these areas. Summary and conclusions are given in the third section. A short bibliography of related research publications is attached in Appendix A.

2.   RESEARCH PROBLEMS IN CIM NETWORKING

Research in CIM networking should be oriented with the objective of vastly improving productivity and quality in manufacturing for a wide cross-section of industry by eliminating the inefficiency of manual actions in the loop and utilizing human resources outside the manufacturing process for management of the process.

Five major research areas in CIM networking are identified below. Problems requiring basic or applied research are presented in each of these major areas.

2.1.   Research Area 1: Specification and Analysis of Network Protocols

Technologies and tools are needed for development of CIM network protocols to assure transparent communication between heterogeneous manufacturing processes and to allow for exception handling without disrupting any real-time operations. Therefore, protocols are to be specified in view of requirements of the processes that are served by the network, and these protocols must be analyzed to verify whether the specified requirements are indeed satisfied. This is done prior to implementation and testing.

2.1.1.   Specification.

Research needs in the area of specification fall into two broad categories. First, there is a need for good descriptive techniques that can model protocols, environments, and requirements. To study the network protocols in a manufacturing environment, each of these components needs to be specified in a common descriptive language. The second broad area in which there is a need for further research is design or architecture, that is, how to build a description of the integrated system. For example, to study issues of network protocols in a new automated factory, it is necessary to have (a) a model of the factory environment including the requirements for its operations, and (b) a network architecture. Research in the area of protocol specification essentially emanates from the need to develop the formalism for capturing the design and architecture of an integrated manufacturing system.

Since time-critical operations are important in a manufacturing environment, the modeling tools should be able to explicitly refer to time in the model and make decisions based on passage of time. As an embedded part, non-time-critical behavior should also be modeled. The operational aspects of design require more research into the issue of levels of abstraction. It is possible to model the CIM system at several levels and relate these levels formally. That is, the integrated system software can be constructed from more basic components, much as hardware is constructed today.

Many classical disciplines in which research work has been done (or is being done) could be useful in exploring powerful and expressive models. Specification techniques, as used in large-scale computer systems, provide the ability to model parallelism and coordination which are essential for describing integrated manufacturing processes. The area of logic and temporal logic is another example. Discrete event simulation captures one type of specification. A fairly powerful paradigm that could be explored is the condition/action approach from artificial intelligence.

2.1.2. Analysis.

The classical techniques for protocol analysis, such as finite-state modeling, Petri net modeling, and discrete-event simulation are largely independent of the characteristics of any particular network. However, an increasingly common view is that analysis of a system must be integrated with the specifications. That is, there should be a common environment in which both specification and analysis can be carried out. This is a fertile research area for protocol analysis for all types of integrated networks.

First, there is functional or logical analysis. The question one asks is *does it work?* That is, a major problem is simply ensuring that the protocol works. In this regard, one could investigate ways for sanity checking, that is, verifying the specifications that would check the self-consistency of the description. This might be a compiler that could look at the static description of the protocol and decide if there is any inconsistency in the description. Similarly, it should be possible to analyze the protocol description to check if it meets the minimal requirements. Analysis could be iterative in which the model description and analysis tools aid in checking more and more elaborate constraints. Research could be done in the area of proof checking in the sense that a suggested proof that the system meets given requirements could be provided. The analysis tools would then aid in checking if the proposed proof makes sense. This is related to work on liveness and safety in protocols which are critical for reliable communications within a CIM network.

The second component is performance analysis. The question one asks is *how well does it work?* Research in this area would be related to measuring and observing a system that is composed of the network and the manufacturing environment. Generally, performance tools for evaluating a protocol or an architecture consisting of several protocols have relied on external analysis of simulation runs. It should be possible to embed more of the observation task into the description of the model itself. This might also yield insight into how to modify the system description if it does not meet performance constraints. An aspect of research is to see to what extent the performance analysis can be done concurrently with the operation of the system, without incurring any penalties due to the observation.

The third component related to protocol analysis is understanding the impact of exceptions. The tools should be able to treat such incidents as breakdowns and fault conditions. For example, a research problem is to determine where the fault lies once an exception has been raised.

Finally, tools are needed for integrating the results of analysis at various levels of protocols. These tools would also be useful for analyzing critical parts of the integrated network. This is related to being able to describe the specification in a hierarchical fashion. The tools should make it convenient to analyze a given system at several levels of detail.

Appropriate analysis tools depend on the particular specification approach. Once such an approach has been adopted, the area of software engineering offers many insights into better tools. As analysis becomes more complex due to the complexity of the system, more work needs to be done on software that can provide high-quality input/output to the system.

2.2.   Research Area 2: Testing and Standardization of Network Protocols

Testing and evaluation of protocols can be either *model-based* or *implementation-based*. The model-based approach is viewed as an extension of protocol analysis and can be carried out as a continuation. The implementation-based approach is essentially product-oriented and has the following test objectives.

Interoperability: This ensures that the product indeed interoperates with the manufacturing environment.

Conformance: This ensures that the product conforms to the specification, thereby increasing the confidence in interoperability.

Performance: This ensures that the product meets the expected level of performance.

Research in this area is *applied* in nature and is essential for standardization of protocols. Topics are discussed below.

2.2.1.   Interoperability Testing.

Success of many critical operations such as exception handling and distributed database management is contingent upon interoperability between host processes in the network. Since interoperability is not defined precisely, it cannot be quantified by developing an accurate measure. Also since it is not a transitive property, each potential pair of processes that need to communicate with each other must be tested for interoperability.

Complete testing for interoperability in a large manufacturing system may not be possible. Development of analytical methods for evaluating interoperability and its testability is a major research area.

2.2.2.   Conformance Testing.

The concept of conformance is closely related to that of interoperability but conformance testing procedures are more complex than those for interoperability. A major problem in conformance testing is to develop methodologies for handling exceptions. There is also a need for fundamental research in the area of provability and verifiability of logical properties of the interoperability and conformance test tools.

Since conventional techniques like finite-state machine modeling may not be applicable to interoperability and conformance testing, a probabilistic approach appears to be a good choice. This is particularly true for an integrated network system.

2.2.3.   Performance Analysis of Integrated Networks.

The performance criteria for integrated network systems vary with different user requirements and viewpoints, for example, functional requirements at the system level and equipment level are significantly different. Typical research areas for network performance analysis are modeling of workload and control variables, selection of configuration variables and parameters for MAP/TOP modules, and optimal design of ISO/OSI functional modules.

Data distribution among heterogeneous sources in the CIM environment has a major impact on network loading. Modeling of network loading requires decomposition strategies for defining basic network management and control functions. Since the structure of function/data modules could significantly influence network performance, research efforts should be focused on topological decomposition of the network into subnetworks.

2.2.4. MAP and TOP Specifications.

The future MAP specifications could be largely influenced by the outcome of research in the following areas: (a) redundancy for enhanced system reliability, (b) capability to meet the needs for CIM evolution, (c) priority options and scheduling of tasks within protocol modules, and (d) real-time operations.

Implementation of TOP in transaction processing in manufacturing environments is a general research area. Examples of specific problems are distributed/remote database access, forms mode in virtual terminal protocols, and network security.

MAP and TOP have similar architectures with the exception of the medium access control (MAC) layer and the application layer. The MAC layers protocols are designed to be different because of the different traffic characteristics in the factory and office environments. The application layer protocols in MAP and TOP are discussed under *user services.*

A potential problem in interfacing MAP with TOP lies in the exact implementation of the two MAC protocols (developed by different manufacturers) at the bridge. This difficulty can be circumvented by a hybrid link-layer protocol which does not require any bridges for interconnecting factory and office communication networks. The research in this category will enhance integration of MAP and TOP with other standard (such as SNA and DECNET) and nonstandard architectures.

Fiber optic communication media with very high bandwidth capacity could accommodate heterogeneous network traffic such as data, voice, video, and facsimile to be transmitted over the same medium. This justifies research for improving the existing MAC layer protocols in MAP and TOP.

2.2.5. User Services.

It is generally agreed that user services are a major research issue in the development of application layer protocols. Simple software links may not be adequate for accessing these protocols but object-oriented languages could be very beneficial. In that case, appropriate compilers must be developed to conform with the needs of specific CIM applications. This is deemed essential for handling distributed database systems.

The role of forms made within the virtual terminal protocol is an important issue as future extensions beyond the current languages and graphics, as used in TOP and MAP, may load the imaging specifications. There is also a need for developing a common language at the equipment level since the relay ladder logic used in programmable logic controllers (PLCs) is often the only way to interface to heterogeneous equipments.

2.2.6.   Human Factors.

Industry is faced with ever increasing complexity of equipment with diminishing human resources to manage and maintain automated systems. Researchers should consider human factors, training of personnel, and organizational design.

2.3.   Research Area 3: Lightwave and Wireless Communication Media

Local wireless communication in manufacturing plants is necessary when terminals are mobile. Problems that arise when radio communication is used for local communication in plants are caused by the multiple access and broadcast properties of the medium as well as by its regulated nature. These problems manifest themselves as severe degradation of the quality of the link (in terms of bit-error-rate, bandwidth, delay, power, etc.), and are caused by multipath and other user interferences. Techniques used to overcome these problems include:

1. Spread spectrum signaling
2. Infrared communication
3. Antenna diversity
4. Improved modulation and pulse shaping
5. Dynamic assignment of the communication channel capacity.

It turns out that the architectural and functional design of a manufacturing plant can have a significant effect on the quality of the wireless communication that can be achieved. The building structure affects the *radio communications* in a manner similar to *theater acoustics*. The spatial configuration of the terminals used in the manufacturing process can also have a dramatic effect on the interference levels. Finally, care in the job scheduling can considerably reduce the maximum bit rate needed for interterminal communications.

The main advantage of wireless communication is mobility. Both radio and infrared transmission are currently used for wireless communication in manufacturing applications but interference and multipath effects limit the scope of applications. The key research issues in mobile communications for manufacturing are:

2.3.1.   Radio Communications in Manufacturing Environment.

The initial factory layout can be adapted to the characteristics of radio and infrared transmission in order to reduce shadow areas where transmission is poor. In a closed environment, attenuation and fading can be reduced appreciably if the factory architecture is properly chosen. Research is needed to develop architectural design rules which will minimize the effects of architecture on radio and infrared transmission. In addition, location of factory machinery can have major effects. Factory layout recommendations with quantitative predictions of multipath, attenuation, and fading characteristics are needed.

2.3.2. Incorporation of Communication Issues into Manufacturing Architecture.

Modeling is necessary for characterizing the properties of statistical channels. It also provides measures of the effects of electromagnetic interference due to stationary components like machine tools and computers, and mobile components like robots, automated guided vehicles, and human beings.

An important research issue is interference effects resulting from multiple accessing by mobile users. Protocols that are currently used for accessing networks are designed for general communication systems and do not consider the special requirements of manufacturing applications.

Another important research topic is determination of modulation and coding techniques to allow for bit-rate increase and reduction of interference and multipath phenomena.

Lightwave communication using optical fiber as the transmission medium has many characteristics which are well-suited to the factory environment. The fiber medium has low transmission loss, is resistant to electromagnetic interference, resists corrosion, and is easily maintained. In addition, since it does not carry current, it would not create sparks in explosive environments and does not cause ground-loop problems.

The characteristics of electro-optic components used in conjunction with fibers impose certain limitations on the network architectures to be implemented in a factory environment. Some of these restrictions are fundamental to the medium while others are transitory and are due to the state of development of the optical components.

2.3.3. Fiber-optics in Manufacturing Systems.

Continuing research is needed into the topologies best suited for use in manufacturing system networks because the characteristics of available electro-optic components are changing so rapidly that optimum topologies must be reevaluated regularly. The relative performance of star, ring, and bus topologies in manufacturing systems is also strongly affected by the medium access protocols. The access protocol used in MAP does not apply the available high fiber optic data rate efficiently and therefore new access protocols need to be considered. In addition to high data rates, these access protocols might also require a different topology to work most efficiently. In particular, improved lower layer protocols are needed to provide adequate real-time performance.

2.3.4. Considerations for CIM Networking.

The optimum use of computer networks in manufacturing systems will require an interdisciplinary approach in order to optimize the interaction of manufacturing process components with message volume and communication connectivity. It is important to properly include the effects of lower layer protocols on the delivery of upper layer services. This means that chemical, electrical, industrial, and mechanical engineers' expertises should be combined in research on issues which involve finding cost-effective tradeoffs among protocol layers for network communication in factory applications.

2.4.   Research Area 4: Real-Time Communications, Control, and Fault Tolerance

Although communication networks involving multiple computers offer enhanced flexibility in real-time distributed control of manufacturing operations, they are accompanied by induced complexities like data latency, mis-synchronization, loss of data due to network failures, noise contamination and buffer saturation, and so on. Product quality and operational safety in manufacturing systems could be seriously degraded unless these phenomena are adequately understood and appropriate actions are taken. Analytical and experimental research is needed to resolve these problems for manufacturing system automation.

2.4.1.   Network-Induced Delays.

In distributed data communication and control systems, network-induced delays occur in addition to the sampling and data processing delays that are inherent in all digital control systems. The impact of network-induced delays on dynamic performance and robustness of real-time control systems need to be investigated for computer-integrated manufacturing.

Data latency in computer networks is time-varying and possibly stochastic as it is dependent on the intensity and distribution of network traffic and specific characteristics of the protocol. Therefore, conventional frequency domain techniques that are used for linear time-invariant systems are not applicable to these systems which are subjected to time-varying delays. Under the present state-of-the-art, combined discrete-event and continuous-time simulation techniques are suitable for understanding and verifying the characteristics of network-induced delays as well as their impact on performance and stability of control systems. Techniques for interpolating and extrapolating useful information from simulation results at selected operating points need to be developed. An example is the perturbation technique for discrete-event system simulation.

In addition to the simulation tools, rigorous analytical techniques are required for analysis and design of real-time control systems that are subjected to network-induced delays because extensive simulation runs could be expensive and may not prove to be the exhaustive means for assuring stability and robustness. Application of advanced analytical techniques such as those cited in Appendix A should be investigated.

2.4.2.   Time Synchronization of System Components.

Time-varying data latency coupled with mis-synchronization between system components could have a devastating effect on both performance and stability of real-time control systems. It is important to note that this synchronization is between the network system components and is, therefore, relatively loose compared to that required for internal operations of tightly coupled computer systems.

In order to circumvent the above problem of mis-synchronization between system components, several methods such as time synchronization and time-stamping based on global clocks could be applied. However, in distributed real-time applications, it may not be easy to implement and maintain accuracy of global clocks in physically dispersed system components. An alternative is periodic transmission of the synchronization message which carries a special bit pattern instead of time values. As soon as a synchronization message is received, system components reset their clock to a predetermined value. Effectiveness and reliability of existing methods should be investigated along with the development of new techniques.

2.4.3.   Fault Tolerance.

Both communication and control systems must be fault-tolerant as their functions are critical for real-time operations in a manufacturing environment. Efficient and reliable methods for fault detection, isolation, and reconfiguration (FDIR) should be studied by considering various aspects such as the degree of fault-tolerant functions. Fault diagnosis implies not only finding the erroneous component but also determining the cause of error. Isolation of the faulty components should not affect the performance of other normally functioning components, and must allow the system to operate in a gracefully degraded mode. Error recovery/reconfiguration may be achieved by time redundancy and/or space redundancy (e.g., by installing redundant buses, network terminals, and processors).

2.4.4.   Task Assignment and Message Scheduling.

These are critical in real-time systems. Each task should be properly scheduled so that an intra-workcell process completes its operations in time. Also the message should be scheduled such that the throughput is maximized and the data latency is minimized. Different workcells may have different message scheduling patterns according to their performance characteristics. For example, an inner control loop with fast dynamics has to sample their data fast. On the other hand, a control loop which has slower dynamics does not have to sample its data frequently. Since the network environment is not static, message scheduling and task assignment should be performed dynamically. These problems may not have direct analytical solutions and it is usually difficult to solve them numerically; therefore heuristic approaches need to be developed.

2.4.5.   Message Prioritization.

Individual workcells may have different priorities which could vary with time according to their functional characteristics. On the other hand, station management messages such as a synchronization message or a fault-recovery message must be transmitted as soon as possible and should have a higher priority than others. However, the notion of a priority scheme is not well defined for real-time applications partly because there are not many systems in operation and the impact of priority of a message upon the system performance is not well understood. Therefore, research efforts should be devoted to find out what difference a priority scheme can make.

2.4.6. Real-Time Programming Languages.

One of the key aspects for networking in the manufacturing environment is the ability to model time appropriately. As time-critical operations are very important, the meaning of real time needs to be defined in a formal way. Thus, standard programming languages that do not have the capability of modeling time explicitly may not be suitable. On the other hand, many operations in this environment might not care about real time, and so everything should not necessarily be slaved to real-time clock(s). Periodic operations, such as those in an assembly line, are an important component in manufacturing, and this again relates to how time is handled. Similarly, the notion of priorities and being able to interrupt an ongoing operation is a question of properly handling the *real-time* problem.

2.5.   Research Area 5: Network Management and Control

The research topics in management and control of CIM networks are discussed below.

2.5.1. Network Control Center (NCC) Architecture.

A critical factor in maintaining control of automated manufacturing processes is its ability to support real-time communications between physically dispersed and functionally distributed elements including engineering design, product distribution, and customer interface. Network management and control, in this context, provide dynamic monitoring and allocation of available communication resources among competing heterogeneous nodes. The CIM network manager should be aware of the status of these resources and, on the basis of this information, must make appropriate decisions. The policies that these decisions implement are derived from the design specifications of the equipments, performance criteria, characteristics of the users' requirements, and the real-time operational environment.

In the dynamic environment of flexible manufacturing systems, the same robot, machine tool, or intelligent CIM terminal may be assigned to different manufacturing processes. The common equipment could be interconnected by a reliable high-speed network which facilitates dynamic reallocation of resources. A centralized network control architecture for this environment may not be cost-effective. On the other hand, a totally distributed NCC may itself generate enough traffic to degrade the overall performance of the network. A suitably distributed network control architecture needs to be developed as an optimal choice for the CIM environment.

2.5.2. Fault Management.

Managing faults is an essential function of manufacturing system networks. In the CIM environment, the network must be designed to communicate the failure information for both manufacturing components and network elements. The management and control of a large network from a remote control center is an area that incorporates both fault analyses and quality control. This is especially important in an integrated production environment where one failure can cause chain reactions of delay and congestion resulting in loss of control of safety and productivity. Characterization of instabilities in manufacturing networks is recommended as a major research area which would lead to realistic error models.

The traditional AI approach to fault diagnosis is not always effective in automation tasks, even at a preliminary level, which rely on dissimilar sensory activities. A viable alternative is the neural network [5,6] which is rapidly gaining importance in diverse disciplines and particularly in situations that require multisensor data fusion [7]. A neural network is a parallel processing architecture where knowledge is represented not only algorithmically in individual processors but also in terms of connections between these processing elements. A neural network represents physical processes with many different inputs, allows use of adjustable parameters in modeling, and is not restricted to the use of models based on overly simplified assumptions.

Two-dimensional signal processing has recently received much attention through emergence of the machine vision technology. In the CIM environment, information from diverse sensor systems can be synthesized via a neural network for prediction, classification, pattern recognition, image understanding, and signal interpretation. Such an automated system can partially emulate the thought process of an experienced human operator for inspection, diagnostics and process control, and also be used as a training tool. Research in formulation and resolution of these problems, in terms of neural network implementations, has a potential for enhancing quality, flexibility, and production rate of CIM processes.

2.5.3. Performance Management.

Performance measures for manufacturing networks are not well characterized. Identification of these performance parameters along with other typical parameters like throughput and data latency is essential for establishing acceptable network performance criteria on the factory floor. The network control center design can then be enhanced to monitor performance and bring performance improvement activities online whenever the performance degrades to unacceptable levels.

2.5.4. Adaptive Resource Management.

Flexibility in a fully automated manufacturing plant may often require physical or logical reconfiguration of the integrated networks. Resource modules specific to different manufacturing environments need to be developed for optimal utilization of the factory equipment. A fully automated manufacturing facility is highly dynamic in nature due to rapid changes in network traffic and operational requirements. To this effect systematic research is needed for developing station management functions which could extend from the link layer up to the application layer.

2.5.5. Network Security.

Communications between databases for manufacturing and marketing via a distributed network provide potential access to this information base by any user of the network. Security of this sensitive data is a major concern. Access control, encryption, authentication of the users, and formal verification of software are some of the techniques which have been developed and implemented by defense and commercial networks to deal with the problems of network security. An example is the *Network Security Evaluation Criteria* that are being developed by the National Security Agency. Transferring this technology from the defense and commercial sectors to the manufacturing sector is an important research topic.

2.5.6. Network Operator Issues.

The network control operator in the CIM environment will need real-time access to distributed network control information which might only be available at distinct physical locations in possibly incompatible databases. Also, the operator must respond to and track all the status and alarm information. The complexity of this information, and the need to respond to failures and reconfiguration requests has recently instigated research in the areas of:

- Distributed database access
- Data compression, fusion, and display
- Intelligent Computer-Aided Instructions.

The use of artificial intelligence techniques for design automation, knowledge capture, and self-learning is a fertile area of research. The use of natural language voice interface for access to distributed databases and to input network control commands is a promising area for simplifying human interfaces to a diversified CIM network management and control system.

Another important area of research is the design of the network operator's console which receives and displays status and alarm data from all elements of the network. Rule-based techniques to diagnose faults remotely and to make inferences from correlations in reported alarm conditions can be incorporated in the operator's console to reduce human labor.

3. SUMMARY, CONCLUSIONS, AND RECOMMENDATIONS

Although networking is essential for providing reliable communications between heterogeneous functions in computer-integrated manufacturing (CIM), research in this field is not yet established. This chapter identifies pertinent research problems and discusses research topics as well as provides recommendations for research in several areas of CIM networking. The materials, presented in this chapter, are expected to be useful for enhancing both basic and applied research in computer networking for manufacturing systems.

Problems of CIM networking have certain similarities with those encountered in other applications including defense and commercial communications. Therefore, programs of national importance like battle management, communications architectures for air/space defense, and integration of commercial information services may directly benefit from CIM networking research and vice versa. An efficient approach to achieving the goals of CIM is to (a) formulate the manufacturing system network specifications and select pertinent problems for collaborative research between universities, industry, and Government institutions, and (b) transfer technology from the military and commercial worlds for resolution of common issues.

REFERENCES

1. A. Ray, "Computer Networking for Manufacturing Systems," *Progress in Robotics and Intelligent Systems*, Volume 1. Norwood, NJ: Ablex, 1994, 78–105.
2. "Manufacturing Automation Protocol (MAP) 3.0 Implementation Release," MAP/TOP Users Group, One SME Drive, P.O. Box 930, Dearborn, MI 48121.
3. "Technical and Office Protocols (TOP) 3.0 Implementation Release," MAP/TOP Users Group, One SME Drive, P.O. Box 930, Dearborn, MI 48121.
4. A. Ray and S. Phoha, eds., *Proceedings of the National Science Foundation Workshop on Computer Networking for Manufacturing Systems*, November 4–6, 1987.
5. G.E. Hinton and S.E. Fahlman, "Connectionist Architectures for Artificial Intelligence," *IEEE Computer*, January 1987, pp. 100–109.
6. J. Hopfield and D.W. Tank, "Computing with Neural Circuits: A Model," *Science*, Vol. 233, August 8, 1986, pp. 625–633.
7. S. Rangawala and D.A. Dornfeld, "Integration of Sensors via Neural Networks for Detection of Tool Wear States," *Proceedings of ASME Symposium on Integrated and Intelligent Manufacturing: Analysis and Synthesis*, New York, 1987.

APPENDIX A. BIBLIOGRAPHY OF RESEARCH MATERIALS IN CIM NETWORKING

Research Area 1: Specification and Analysis of Network Protocols

A.V. Aho, J.E. Hopcroft and J.D. Ullman, *The Design and Analysis of Computer Algorithms*. Reading, MA: Addison-Wesley, 1985.

C.G. Cassandras and S.G. Strickland, "Perturbation Analytic Methodologies for Design and Optimization of Communication Networks," *IEEE Journal of Selected Areas in Communications*, Vol. 6, No. 1, January 1988, pp. 158–171.

C.A. Sunshine (ed.), *Communication Protocol Modeling,* Dedham, MA: ARTECH House, 1981.

P.E. Green (ed.), *Computer Networks and Protocols,* New York: Plenum, 1983.

Y-C. Ho, "Performance evaluation and perturbation analysis of discrete-event dynamic systems," *IEEE Trans. on Aut. Contr.*, Vol. AC-32, No. 7, July 1987.

C.A. Sunshine (ed.), *Protocol, Specification, Testing, and Verification II.* Amsterdam: North Holland, 1982.

H. Rudin and C. West (eds.), *Protocol Specification, Testing, and Verification III.* Amsterdam: North Holland, 1983.

G.V. Bochmann and B. Sarikaya (eds.), *Protocol Specification, Testing, and Verification VI.* Amsterdam: North Holland, 1986.

C.A. Sunshine, "Formal Techniques for Protocol Specification Method," *IEEE Computer,* September 1979, pp. 20–27.

R. Rosenthal (ed.), *Workshop on Factory Communications,* U.S. Department of Commerce Special Publication No. NBSIR 87-3516, March 1987.

Research Area 2: Testing and Standardization of Network Protocols

E. Barkmeyer et al., *An Architecture for Distributed Data Management in Computer Integrated Manufacturing,* U.S. Department of Commerce, Publication No. NBSIR 86-3312, April 1986.

J.A. Daigle, A. Seidmann and J. Pimentel (eds.), *IEEE Network Magazine: Special Issue on Communications for Manufacturing,* May 1988.

Implementation Agreements for Open Systems Interconnection Protocols, U.S. Department of Commerce Publication No. NBSIR 86-3385-2, October 1986.

R.S. Matthews, K.H. Muralidhar and S.R. Sparks, "MAP 2.1 Conformance Testing Tools," *IEEE Trans. Software Eng.,* March 1988.

R. Rosenthal (ed.), *Workshop on Factory Communications,* U.S. Department of Commerce Special Publication No. NBSIR 87-3516, March 1987.

Research Area 3: Lightwave and Wireless Communication Media

IEEE Communications Magazine: Mobile Radio Communications, Vol. 24, No. 2, February 1986.

IEEE Communications Magazine: Microwave Digital Radio, Vol. 24, No. 8, August 1986.

IEEE Communications Magazine: Mobile Communications, Vol. 25, No. 6, June 1987.

IEEE Communications Magazine: Echo and Delay in Digital Mobile Radio, Vol. 25, No. 8, August 1987.

IEEE Communications Magazine: Terrastial Optical Fiber Systems, Vol. 25, No. 10, October 1987.

H. Ohnsorge (ed.), *IEEE Journal on Selected Areas in Communications: Broadband Communications Systems*, Vol. SAC-4, No. 4, July 1986.

L.J. Scerbo, J.P. Varachi, H. Murata and P.L. Pope (eds.), *IEEE Journal on Selected Areas in Communications: Engineering and Field Experience with Fiber Optic Systems*, Vol. SAC-4, No. 5, August 1986.

D.C. Gloge, R.C. Mendez, J.E. Midwinter, S.T. Personick and S.D. Shimada (eds.), *IEEE Journal on Selected Areas in Communications: Special Issue on Fiber Optic Systems for Terrastial Applications*, Vol. SAC-4, No. 9, December 1986.

K. Brayer (ed.), *IEEE Journal on Selected Areas in Communications: Special Issue on Fading and Multipath Channel Communications*, Vol. SAC-5, No. 2, February 1987.

J.C.Y. Huang, K. Kohiyama, A. Leclert and F. Siebelink (eds.), *IEEE Journal on Selected Areas in Communications: Advances in Digital Communications by Radio*, Vol. SAC-5, No. 3, April 1987.

D.C. Cox, K. Hirade and S.A. Mahmoud (eds.), *IEEE Journal on Selected Areas in Communications: Special Issue on Portable and Mobile Communications*, Vol. SAC-5, No. 5, June 1987.

IEEE Network Magazine: Special Issue on Communications for Manufacturing, Vol. 2, No. 3, May 1988.

A. Ayyagari and A. Ray, "A Fiber-Optic-Based Protocol for Manufacturing System Networks—Part I—Conceptual Development and Architecture," *ASME Journal of Dynamic Systems, Measurement and Control*, Vol. 114, No. 1, March 1992, pp. 114–120.

A. Ayyagari and A. Ray, "A Fiber-Optic-Based Protocol for Manufacturing System Networks—Part II—Statistical Analysis," *ASME Journal of Dynamic Systems, Measurement and Control*, Vol. 114, No. 1, March 1992, pp. 121–131.

Research Area 4: Real-Time Communications, Control, and Fault Tolerance

A.P. Belle Isle, "Stability of Systems with Nonlinear Feedback Through Randomly Time-Varying Delays," *IEEE Trans. on Automatic Control*, Vol. AC-20, No. 1, February 1975, pp. 67–75.

K. Hirai and Y. Satoh, "Stability of a System with Variable Time Delay," *IEEE Trans. on Automatic Control*, Vol. AC-25, No. 3, June 1980, pp. 552–554.

Y-C. Ho and X-R. Cao, "Performance Sensitivity to Routing Changes in Queueing Networks and Flexible Manufacturing systems Using Perturbation Analysis," *IEEE Journal of Robotics and Automation*, Vol. RA-1, No. 4, December 1985, pp. 165–172.

Y-C. Ho, "Performance Evaluation and Perturbation Analysis of Discrete-Event Dynamic Systems," *IEEE Trans. on Aut. Contr.*, Vol. AC-32, No. 7, July 1987, pp. 563–572.

F. Kozin, "A Survey of Stability of Stochastic Systems," *Automatica*, Vol. 5, 1969, pp. 95–112.

R. Luck and A. Ray, "An Observer-based Compensator for Distributed Delays," *Automatica*, Vol. 26, No. 5, 1990, pp. 903–908.

A. Ray, "Performance Evaluation of Medium Access Control Protocols for Distributed Digital Avionics," *ASME Journal of Dynamic Systems, Measurement and Control*, Vol. 109, No. 4, December 1987, pp. 370–377.

A. Ray, "Distributed Data Communication Networks for Real-Time Process Control," *Chemical Engineering Communications*, Vol. 65, March 1988, pp. 139–154.

A. Ray, "Networking for Computer-Integrated Manufacturing," *IEEE Network Magazine: Special Issue on Communications for Manufacturing*, Vol. 2, No. 3, May 1988, pp. 40–47.

Y. Halevi and A. Ray, "Integrated Communication and Control Systems: Part I—Analysis," *ASME Journal of Dynamic Systems, Measurement and Control*, December 1988, pp. 367–373.

A. Ray and Y. Halevi, "Integrated Communication and Control Systems: Part II—Design Considerations," *ASME Journal of Dynamic Systems, Measurement and Control*, December 1988, 374–381.

A. Ray, L.W. Lion and J.H. Shen, "State Estimation Using Randomly Delayed Measurements," *ASME Journal of Dynamic Systems, Measurement and Control*, Vol. 115, No. 1, March 1993, pp. 19–26.

J.H. Shen and A. Ray, "Extended Discrete-time LTR Synthesis of Delayed Control Systems," *Automatica*, Vol. 29, No. 2, pp. 431–438, 1991.

L.W. Lion and A. Ray, "A Stochastic Regulator for Integrated Communication and Control Systems: Part I—Formulation of Control Law," *ASME Journal of Dynamic Systems, Measurement and Control*, Vol. 113, No. 4, December 1991, pp. 604–611.

L.W. Lion and A. Ray, "A Stochastic Regulator for Integrated Communication and Control Systems: Part II—Numerical Analysis and Simulation," *ASME Journal of Dynamic Systems, Measurement and Control*, Vol. 113, No. 4, December 1991, pp. 612–619.

R. Luck, A. Ray and Y. Halevi, "Observability under Recurrent Loss of Data," *AIAA Journal of Guidance, Control and Dynamics*, Vol. 15, No. 1, January–February 1992, pp. 284–287.

A. Ray, "Output Feedback Control Under Randomly Varying Distributed Delays," *AIAA Journal of Guidance, Control and Dynamics*, Vol. 17, No. 4, July–August 1994, pp. 701–711.

U. Rembold, C. Blume and R. Dillman, *Computer-Integrated Manufacturing Technology and Systems*. New York: Marcel Decker, 1985.

K.G. Shin, C.M. Krishna and Y.H. Lee, "A Unified Method for Evaluating Real-Time Computer Controllers and Its Applications," *IEEE Trans. on Automatic Control*, Vol. AC-30, No. 4, April 1985, pp. 357–366.

Research Area 5: Network Management and Control

D. Bertsekas and R. Gallager, *Data Networks*. Englewood Cliffs, NJ: Prentice Hall, 1987.

F. Hayes-Roth, D.A. Waterman and D.B. Lenat, *Building Expert Systems*. Reading, MA: Addison-Wesley, 1983.

J. Hopefield, "Neural Networks and Physical Systems with Emergent Collective Computational Abilities," *Proc. Natl. Academy of Sciences,* Vol. 79, April 1982, pp. 2554–2558.

IEEE Communications Magazine: Network Planning, Vol. 25, No. 9, September 1987.

IEEE Communications Magazine: Artificial Intelligence Communications, Vol. 26, No. 3, March 1988.

H.A. Malec (ed.), *IEEE Journal on Selected Areas in Communications: Quality Assurance for the Communications Community,* Vol. SAC-4, No. 7, October 1986.

C. Dhas, W. Montgomery, L. Stringa and V. Konangi (eds.), *IEEE Journal on Selected Areas in Communications: Knowledge-Based Systems for Communications,* Vol. 6, No. 5, June 1988.

IEEE Network Magazine: Special Issue on Network Security, Vol. 1, No. 2, April 1987.

IEEE Network Magazine: Special Issue on Network Management Protocols, Vol. 2, No. 2, March 1988.

IEEE Network Magazine: Special Issue on Expert Systems in Network Management, Vol. 2, No. 5, September 1988.

Implementation Agreements for Open Systems Interconnection Protocols, U.S. Department of Commerce Publication No. NBSIR 86-3385-2, October 1986.

S.D. Kuzmok, M.D. Brittingham, A.L. Gorin, G.A. Millich and J.E. Shoenfelt, "Knowledge-Based Signal Interpretation," *AT&T Technical Journal,* Vol. 67, Issue 1, January/February 1988, pp. 104–120.

D. Rumelhart and J. McClelland, *Parallel Distributed Processing,* Vol. 1. Cambridge, MA: MIT Press, 1986.

M.G. Wihl, "Neural Networks Make Cybernetic Multisensor Systems Possible," *Journal of Machine Perception,* September 1988, pp. 11–18.

P.K. Wright and D.A. Bourne, *Manufacturing Intelligence*. Reading, MA: Addison-Wesley, 1988.

S. Lee and A. Ray, "Performance Management of Multiple Access Communication Networks," *IEEE Journal of Selected Areas in Communications,* Vol. 11, No. 9, December 1993, pp. 1426–1437.

8

Reduced Protocol Architecture for Factory Applications

Luigi Ciminiera, Claudio Demartini, and Adriano Valenzano
CENS and Dipartimento di Automatica e Informatica
Politecnico di Torino, Italy

1. INTRODUCTION

The main goal of flexible automation is that of going beyond the limits due to the rigid operations carried out by independent groups of robots and automatic devices handled by autonomous processing units, in order to accomplish complete software and hardware integration and to coordinate all the activities supported by the various equipment making up a production system.

It is not possible, today, to assert that the scenario previously mentioned is already operating inside the factory environment. In fact, since automation has been introduced in the manufacturing environment using a "bottom-up" approach, solutions currently developed are based on "automation isles" which are unable to communicate with all levels. Transfers of information are often performed by the line operators which spend a large amount of time hand-carrying data produced in the various working phases between incompatible computer stations, so that successive operators in the production line can read, interpret the data, and then manually insert other data and eventually transfer the result to the next station.

The optimal solution for the communication problem is based on the total integration of the automatic devices in a single distributed information system, which allows data to be transferred from one working process to the next without involving the operator. Furthermore, such a solution would make it possible to update the configuration of the production process in real time modifying some variable parameters such as overproduction, varied market needs, design changes, and product customizing.

An integrated approach to communication issues forces system designers to face the following problems:

- devices and computers used for the organization of a production system are heterogeneous

- environmental conditions, in which control devices and processing units must work, seriously affect the choice of a suitable network topology as well as the physical communication media
- the total number of units making up the distributed system directly affects the response times of the system
- processing times represent, for networks oriented to process control, a basic factor in order to support efficient handling of the tasks assigned to the whole integrated production system
- activities in the factory environment must be coordinated with activities carried out in the office network systems
- installation costs of communication media must be considerably reduced or optimized
- previous investments must be maintained, planning a gradual migration from current implementations to advanced solutions and architectures which, however, must be able to interact with preexistent systems in the factory environment.

Recent advances in the local area network field offer a great variety of solutions which are appropriate for simple physical interconnection requirements as well as for communication protocol definition and development. Many solutions have been proposed and implemented so that users quickly have felt the need to define some reference standards in order to guarantee a higher degree of security and stability in the interconnection rules which have to be adopted in the factory environment. Since the 1970s the manufacturers of robots, machine tools, programmable logic controllers, and numerical controllers have faced the problem, initially proposing independent and partial solutions, based on the availability of customized interfaces so that data could be exchanged with other end-systems provided by different vendors. Such solutions, which were mainly the result of divergent approaches and did not have a total and unique strategy, could not solve the problems related to the implementation costs and efficiency needs due to the ever increasing requirements of integration within the factory applications.

A network standard based on an open architecture is a considerable step towards the solution of many interconnection problems. The OSI (Open Systems Interconnection) model provides a conceptual and functional framework on which to graft different operating systems, different vendors devices, and different media organized onto a single network system. It acknowledges existing protocols by placing them within the model architecture, which also serves as a focal point where all activities related to the development of standards can be coordinated. An OSI network does not imply a specific protocol implementation, but rather a set of standards that various systems can use to exchange information between networks.

1.1. MAP Architecture and Protocol Profiles

In 1981, General Motors, forced by the enormous costs of the solutions implemented to connect automatic devices and computers in its production structures, started developing a specification for a communication protocol suited to industrial automation [1]. This approach was first directed to the GM suppliers, and then rapidly spread out to include the biggest American and European companies, which now actively participate in the discussion and definition of specifications and are grouped in two organizations: MUG-MAP User Group and EMUG-European MAP User Group. The main points which the specifications are based on can be given as follows:

- the reference architecture for MAP protocols is the OSI (Open Systems Interconnection) model standardized by ISO [2,3]
- services provided by protocol levels of the chosen standards must match the application requirements required in the automation and control process environment
- services and protocols of each layer in the architecture must be approved by ISO and NBS (National Bureau of Standards)
- if a standard with specific requirements is not available, then its definition can be completed in the framework of the standard organizations by the constitution of appropriate committees and working groups.

The guidelines of the MAP specifications are intended to define a Local Area Network for terminals, computing resources, and programmable devices within a plant. The architecture also allows for the interconnection of many LANs and for their connection to Wide Area Networks or digital PBXs for long-distance communications.

Figure 8.1 reports the structure of the full MAP specifications version 3.0, and points out the close relation between the protocol architecture and the OSI reference model.

The full MAP is based on a central communication medium named "backbone," based on a 10 Mbps broadband signaling technique which supports many communication channels on the same physical medium. This means that several networks can be run simultaneously by allocating different send and receive channels to each network, and other channels can be used for voice and video communication. The access mechanism is based on the token passing approach. The token represents the right to use the medium, and is offered to each node in turn. The advantage of this approach is that the maximum access time can be calculated and guaranteed, an important feature for process control applications and automation. In order to meet low-cost solution requirements for the interconnection of simple devices, a carrierband solution has also been defined. It is a single channel, lower-cost, slower (5 Mbps) version of the full MAP.

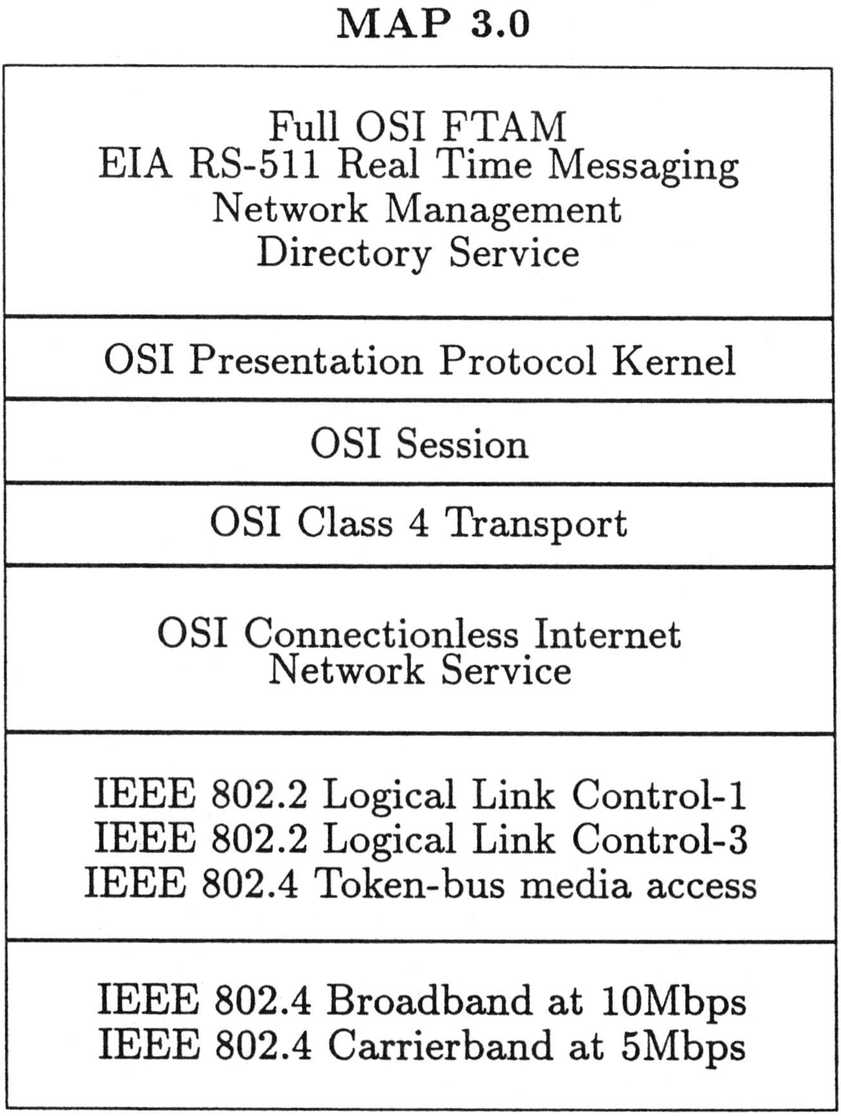

MAP 3.0

Full OSI FTAM EIA RS-511 Real Time Messaging Network Management Directory Service
OSI Presentation Protocol Kernel
OSI Session
OSI Class 4 Transport
OSI Connectionless Internet Network Service
IEEE 802.2 Logical Link Control-1 IEEE 802.2 Logical Link Control-3 IEEE 802.4 Token-bus media access
IEEE 802.4 Broadband at 10Mbps IEEE 802.4 Carrierband at 5Mbps

FIGURE 8.1.   Full MAP protocols profile.

In addition to broadband and carrierband, other devices have been defined and specified to be used within a MAP system to link MAP to MAP or MAP to non-MAP systems or to cope with time-critical applications:

• MAP bridge
• MAP router

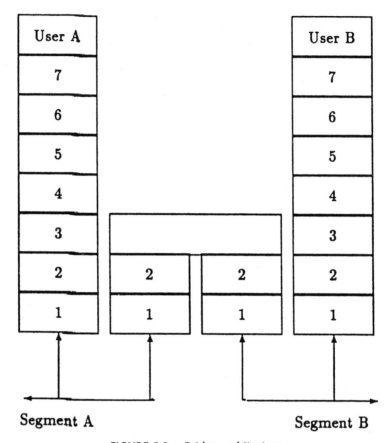

FIGURE 8.2.   Bridge architecture.

- Gateway
- MAP/EPA nodes
- MiniMAP.

1.1.1.   MAP Bridge.

A bridge allows two identical ISO 8802 networks to be connected transparently, which means that it can be ignored by a transmitting node. A store-and-forward facility is also provided. A bridge-based solution allows segmentation to be implemented in order to overcome limitations on network distance and capacity, and faults can be restricted to a single segment without involving other parts of the network. Figure 8.2 shows a scheme for a connection between two segments based on the bridge solution.

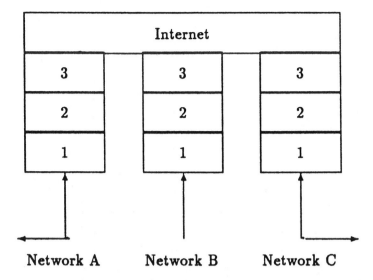

FIGURE 8.3. Router architecture.

1.2. MAP Router

A router connects two or more distinct networks to a common point and is composed of the lower three layers of the MAP architecture as exemplified in Figure 8.3.

It is not transparent to network operations, so that it must be directly addressed when it is used. The router provides routing facilities such as finding paths between different networks having different address domains and frame size. It can be used to link a MAP network to another MAP network or a MAP to a TOP (Technical and Office Protocol) network.

1.3. Gateway

Gateways operate as protocol translators between different networks (whether they are ISO compatible or not). Figure 8.4 shows the gateway structure.

A typical example is represented by the requirement in a manufacturing plant to connect a proprietary subnetwork to the MAP backbone. It must be pointed out that, given the large variety of proprietary networks, gateways are not standard devices and they represent the most expensive interface in a MAP system.

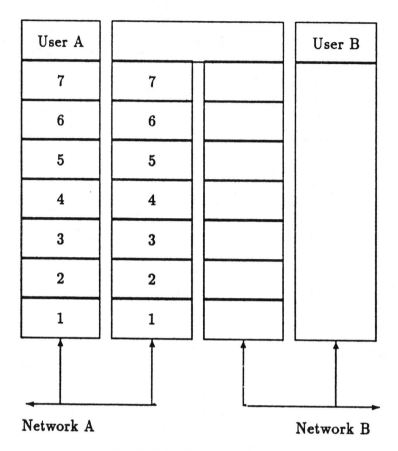

FIGURE 8.4.   Gateway architecture.

1.4.   MiniMAP

The full-MAP architecture provides multiple high-bandwidth noninterfering signal channels. However, the high cost, slow processing time, and large amount of protocol overhead make such a protocol architecture questionable for real-time use among cooperating tasks implementing a cell control.

Hence the solution introduced by the MAP specifications is an architecture based on a reduced protocol profile. Only three layers are supported: Layers 1, 2, and 7 as shown in Figure 8.5.

The node device based on this enhanced performance architecture (EPA) is also called MiniMAP. This solution involves the loss of all those services that in the full architecture are supported by the network, transport, session, and presentation layers. In other words the application layer is directly interfaced to the data link layer.

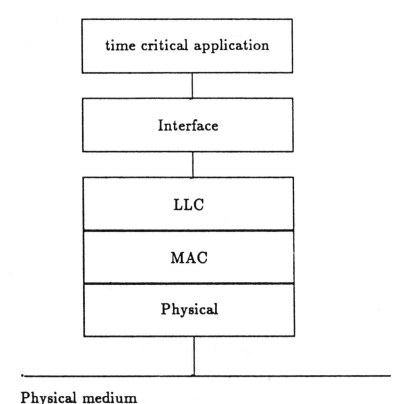

Physical medium

FIGURE 8.5.   MiniMAP protocols profile.

The drawbacks of such an approach are considerable. First of all there is no guarantee of message delivery, which can be confined at the segment level. Furthermore, the message length is confined to the maximum size of the data link unit and only one outstanding message at a time is permitted. Another important aspect is related to presentation services, which can no longer be negotiated but are fixed between the cooperation tasks of the application.

The environment for this architecture relies on the connection of low-cost intelligent devices to a time-critical network linked to a carrier-band MAP by means of a MAP/EPA node.

2.   ACKNOWLEDGED DATAGRAM SERVICE

Eliminating the intermediate layers in the reduced protocol architecture allows faster communication between stations, but also introduces a lack of functionalities assigned in full OSI model to the layers not used in MiniMAP. The characteristics of the subnetworks using the reduced architecture confine the problems introduced by the loss of functionalities. This is particularly so for the fixed network configuration, the a priori knowledge of the functions to be performed, and the extension to a single segment of the network play an important role in the definition of MiniMAP architecture.

Two main approaches have been used to cope with the loss of functionalities due to the unused middle layers: some functionalities have been moved, in a simplified form, to the remaining layers, or some restrictions have been introduced in the communication services. The following is a list of the major functionalities performed by the middle layers with the corresponding solutions adopted for the MAP collapsed architecture.

- *Reliable in-sequence delivery:* The data-link version used in MiniMAP incorporates mechanisms for both acknowledgment generation for the received frames and proper coupling between the received acknowledgments and the data frames to which they are referred; moving these functions to the data link layer is possible because only single segment subnetworks are considered.
- *Routing and network address conversion:* These functions are no longer needed in a single segment subnetwork.
- *Flow control:* This function may be performed by properly setting some parameters of the application layer, that is, such parameters as the maximum number of MMS (Manufacturing Message Specification) associations and the maximum number of outstanding MMS requests for each association.
- *Parameter negotiation:* Most of the unused OSI layers have mechanisms for parameter negotiation; these functions are avoided in MiniMAP by setting fixed values for parameters such as the maximum user packet size (limited to 1 Kbyte) and the session type (full-duplex is allowed only when there is no re-synchronization).
- *Message format definition:* This function, which is performed in the presentation layer by using ASN.1 notation, has been incorporated in the application layer, because the MMS is able to use ASN.1 in the MiniMAP environment, too.

The first type of functions listed above is implemented in MiniMAP by adopting both the IEEE 802 [4] standard and the ISO 8802 [5], with the extensions introduced because of the need to ensure reliable datagram communication [6].

The IEEE 802 standard defines a structure for the data link layer composed by two sublayers: medium access control (MAC) and logical link control (LLC). The former is mainly responsible for controlling the access of the hosts to the channel; the LLC, placed on top of the MAC, provides the tasks using the data link layer services with a uniform interface, which is independent of the medium access protocol.

TABLE 8.1.   Correspondence between classes and types.

		Classes			
		1	2	3	4
	1	X	X	X	X
Types	2		X		X
	3			X	X

Three types of LLC are defined by IEEE 802, and two of them (type 1 and type 3) are used in the MAP 3.0 specifications for the hosts connected to the backbone (type 1) and for those connected to subnetworks (type 3). The main differences between the three types may be summarized as follows:

- **Type 1:** A confirmation for the transmission of a frame is generated for the LLC user as soon as the MAC has notified the transmission of that frame; therefore, the confirmation just means that the frame has been transmitted, while it does not bear any information on regarding the reception of the frame.
- **Type 2:** The frames are transmitted after establishing a connection with the remote station, and using an HDLC-like type of flow control.
- **Type 3:** A confirmation for the transmission of a frame is generated for the LLC user only when an acknowledge frame for the transmitted one is received; in this case, the confirmation also bears information about the frame reception, including the cases of erroneous and nonreception, as this type of LLC includes limited recovery functionalities.

Therefore, Type 3 communications provide the minimum level of functionalities necessary to implement reliable in-sequence communications and thus they may be used in MiniMAP hosts to support the application protocol operations, as in a MiniMAP environment, it is not possible to rely on the checks and recovery mechanisms implemented by the layers 3 to 6 of the full stack of OSI protocols.

A single implementation of the IEEE 802.2 LLC may allow different types of communication, according to the four standard classes defined in Table 8.1. It is possible to see that MiniMAP nodes use both Type 1 and Type 3 communications.

In the rest of this section, the terminology adopted in defining the ISO 8802 standard will be used.

2.1.   Type 3 Services and Sequencing

Basically, the protocol used for Type 3 transmissions is made up of a command frame issued by the station where the initiator user resides; the address of that command frame should provide a response frame, which normally tells the initiator whether or not the command frame has been successfully received, as well as other possible information. The receipt of the response frame closes the service execution, so that any further action will be regarded as nonrelated to the previous ones. Therefore, no connection-oriented environment is maintained, as far as different command-response pairs are considered, so that it is possible to consider the Type 3 services based on a datagram-oriented protocol. Instead when considering each frame transmitted, it is necessary to maintain within the LLC information which is useful to correctly relate the incoming responses to the commands sent, and to avoid frame duplications or losses.

All the Type 3 transmission services are, in general, structured to carry out three types of functions, namely:

- **Request:** passed by the initiator user to the local LLC, it contains all the information required for the activation of the whole service mechanism.
- **Indication:** passed by the responder LLC to the local user when a command frame is received.
- **Status:** passed from the LLC on the initiator station to the local user so as to notify whether or not the service has been completed successfully, and to pass any data received with the response frame to the local user.

2.1.1. Addressing.

Another general characteristic of the IEEE 802 LLC is that the source and destination addresses are service access point (SAP) addresses rather than simple physical addresses. Each SAP may be considered as a bidirectional communication port between the LLC and the user sitting above the data-link layer (note that the user may be the next upper protocol layer). With this addressing convention, each service may address a different remote data-link layer user directly.

This feature is important in the MiniMAP protocol architecture, because MMS may be structured, within each station, in a set of distinct virtual manufacturing devices (VMD), each one implementing all the MMS entities (such as variables, semaphores, program invocations, etc.) visible to the remote application programs and related to a specific real manufacturing device. If each VMD is associated to a different SAP, then the data link addressing scheme provides a straightforward means to address the different VMDs.

The full SAP address is made up of two parts: the local SAP number identifying the SAP within the local LLC entity, and the MAC address, recognized by the MAC sublayer for that LLC entity. The first component will be referred to as SSAP or DSAP according to whether the source or the destination SAP is indicated; similarly, DA and SA will designate the destination and the source MAC addresses. This distinction will be useful when the mechanisms to avoid frame duplication or loss are discussed.

2.1.2. Services.

The main function of this service is to provide the means to reliably transfer data from a user of the initiator LLC to a user of the responding LLC.

This service can be implemented in two ways, depending on the class of service parameter included in the service request. The two methods differ in the data link sublayer generating the response frame, because some types of MAC sublayers are able to respond without LLC intervention; the ability of the different MACs to implement this feature is discussed at the end of this section.

The diagrams showing a normal (error-free) sequence implementing the DL-DATA-ACK service are shown in Figure 8.6 for both MAC-generated response and LLC generated response. In Figure 8.6, the vertical lines show the boundaries between the different sublayers of the initiator and the responder, while the central zone of the diagram represents the physical layer and the channel.

Figure 8.6.a depicts a typical sequence of frames implementing the DL-DATA-ACK service by using a MAC-generated response, as the acknowledge frame is generated without the intervention of the responding LLC. Another important characteristic of this type of implementation is that no other frame is transmitted between the command and the response, as the channel contention resolution mechanism is not restarted until the response has been transmitted. This is not possible for all the types of MACs, hence it is not always possible to implement responses generated by the MAC layer.

In contrast, when the response is generated by the appropriate LLC, the MAC sublayer is not aware that command and response transmissions are two phases of the same operation, thus it restarts the contention resolution mechanism after the transmission of the command. In this way, other frames generated by other stations are allowed to enter on the channel, and the response is transmitted only when the responding station gets the right to transmit on the channel. This is indicated in the Figure 8.6.b with a gap in the vertical lines.

It is interesting to compare Figure 8.6 with Figure 8.7 showing the same time diagram for the DL-DATA service used for Type 1 transmissions. From this comparison it is evident that the status generated by the latter tells the user only that the frame has been transmitted, while it does not bear any information on the successful completion of the transmission, as for the DL-DATA-ACK service.

DL-REPLY. This service is used to transfer data from a remote data-link SAP; the data, transmitted with the response frame, should already be available at the responding data link entity when the command is received, so that no action should be taken by the data link user on the responding side.

The command may also contain data, so that it is possible to implement a fast bidirectional data exchange between two remote data-link SAPs. If the data are not available at the responding side, a response is always generated to tell the initiator that the command has been successfully received. The indication received by the data-link user on the responding side notifies that a DL-REPLY

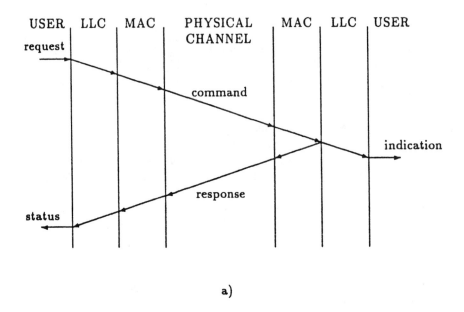

a)

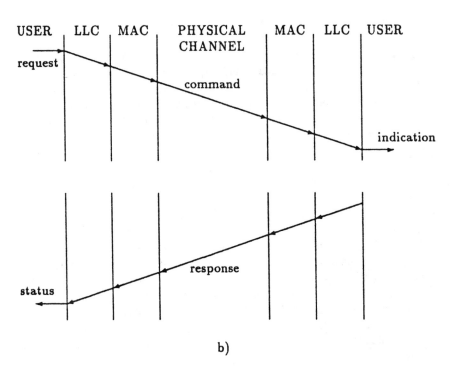

b)

FIGURE 8.6. Time diagram of the implementation of the DL-DATA-ACK service: (a) with MAC generated response; (b) with LLC generated response.

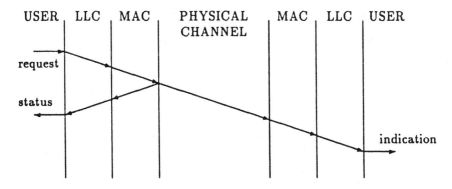

FIGURE 8.7. Time diagram of the implementation of the DL-DATA service.

command has been received, and passes on the data (if any) included in the command frame.

The time diagram showing the operations of this service are identical to those in Figure 8.6; therefore, in this case too it is possible to have a response generated by either the MAC or the LLC sublayer. In the former case, it is possible to implement stations that can only be polled, as they are not able to contend for the channel access; these stations may have cheaper network interfaces, as most of the implementation complexity is due to channel access protocols.

DL-REPLY-UPDATE. This is the only service with only local effects, as it does not cause the transmission of any frame. The DL-REPLY-UPDATE service is used by the data-link user to pass a block of data, which is transmitted as a response to a subsequent DL-REPLY command addressed to the SAP where the DL-REPLY-UPDATE is performed.

The time diagram for this service is shown in Figure 8.8, where only one station is depicted. It is important to note that only one block of data can be pending for each SAP, and its size cannot exceed the maximum size of the data field in a frame.

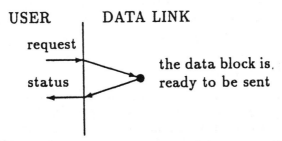

FIGURE 8.8. Time diagram of the implementation of the DL-REPLY-UPDATE service.

2.1.3.   Sequencing.

It has already been pointed out how a data-link layer intended to support reduced protocol profiles should include mechanisms which ensure the correct in-sequence frame reception, with no duplication or loss of frames. This is accomplished in Class 3 LLC by using three types of protocol variables: transmit sequence state variables, receive sequence state variables, and reception status variables.

The first two types of variables are used to generate sequence numbers to be included in the command and response frame, respectively, and to check the values of the incoming response and command frames, as well. They can assume only two values $\{0,1\}$. The reception state variable is used to store the value of the status code transmitted with the last response frame, so that, if the response is lost and the command retransmitted, the value of the status field is always the same as the previous one.

As frames may be exchanged with different stations, it is necessary to recognize which command a particular response is referring to. To do this, one transmit sequence state variable is assigned to each distinct triple (SSAP, DA, priority), where the last element is the MAC priority level used to transmit commands. Similarly, one receive sequence state variable and one reception state variable are associated to each distint triple (SSAP, SA, priority). The result is that only one command frame can be waiting for a response for each distinct triple (SSAP, DA, priority) at any time because the sequence state variables have binary values, hence no particular resequencing mechanism is required.

2.2.   Implementation Issues for Different MAC Protocols

2.2.1.   IEEE 802.3.

Although the CSMA/CD access protocol [7] is not generally considered in the industrial application field, it is included here for sake of completeness.

This type of access protocol allows only LLC-generated responses, as the end of a frame transmission automatically starts the channel contention mechanism; therefore, there is no way of guaranteeing that the response will be the next frame transmitted on the channel after the command frame.

2.2.2.   IEEE 802.4.

The IEEE 802.4 specifications allow for the implementation of an acknowledged datagram service at the data-link level in two different ways. The first solution is to use the immediate acknowledgment mechanism to implement MAC-generated responses. Such a facility is included as an optional service in the 802.4 medium access control (MAC) specifications [8] and is based on the *request with re-*

sponse MAC mechanism. In this case the source MAC service provider sends one user message and then waits for a valid response frame (message acknowledgment) from the destination entity. If the answer is not received within a certain time interval, the message is retransmitted for up to a maximum number of retries. Both transmissions and the possible retries are performed without passing the token; hence no other station is allowed to use the channel during the entire transaction.

The second solution makes use of the traditional LLC-generated frames to send response packets when the token is received by the station where the destination user resides. In other words, the acknowledgment frame is not transmitted by the destination MAC entity when the user message is received; but the addressed station waits for the token to be received before confirming for the sender that the message was correctly received.

2.2.3.   IEEE 802.5.

Although IEEE 802.4 has been selected to be used in MAP, token rings are still important in the industrial application field, as they are the easiest way, at present, to implement fiber optic networks with a large number of stations. This has been acknowledged in the MAP 3.0 specifications, where two alternatives are envisaged for the introduction of fiber optics: IEEE 802.4H for subnetworks with a limited number of stations, and IEEE 802.5 for larger subnetworks.

As is well known the 802.5 token-ring MAC standard specifies a mechanism to acknowledge the reception of a frame. This form of "implicit" acknowledgment makes use of some bits in the frame status field [9] that are set by the receiver and checked by the transmitting station to verify whether or not the message has been correctly obtained by the former.

There are two main reasons why the use of the implicit acknowledgment mechanism to implement a type 3 LLC service over a token ring MAC is discouraged:

1.   the status field in each MAC frame is not covered by the frame check sequence (FCS); hence transmission errors can result either in acknowledgments from incorrect receivers or in losses of correct acknowledgment frames.

2.   it is not possible with this method to implement the MAC generated response for the DL-REPLY service, because in such a case data should be transmitted with the response, in addition to the acknowledgment.

In an IEEE 802.5 system both simple acknowledgments and replies with data can be implemented by LLC frames that are transmitted by the responding station using the standard MAC services once the token has been obtained. Simple acknowledgments are used to implement the responses for the DL-DATA-ACK service, while data in the acknowledgment frames are needed for the DL-REPLY service responses.

3.   IEEE 802.4 PERFORMANCE ANALYSIS

3.1.   Network Models

In this section, evaluation of the performance of token-bus networks which implement the IEEE 802.4 standard and offer an acknowledged datagram service to the data link users is presented.

An important point which must be stressed here is that the acknowledgment mechanisms considered in this chapter concern only the packet delivery functions. In other words, the data-link service performs the packet delivery control for the user message, but does not guarantee the sender that the message is really read and understood by the remote receiving user. The acknowledgment mechanisms only notify the transmitter that the message has correctly reached its destination node.

The two implementation alternatives discussed earlier have been considered in the analysis and modeled in order to evaluate and compare their performances by direct simulation.

The following general assumptions have been made for both systems:

1.   The network architecture includes N stations connected to a shared communication channel based on the IEEE 802.4 specifications [8]: this situation occurs, for instance, when a number of computing devices is able to access the MAP backbone [1] directly.

2.   Each station is modeled by a pair of processes. The first is able to send/receive messages using the communication network, while the second acts as a traffic generator for the whole station since it creates messages and queues them for transmission to the station output buffer.

3.   All user messages have the same length, however different lengths have been used for tokens, acknowledgments, and messages, respectively. Different simulation results have been collected by using the message length as a parameter.

4.   Each message is assigned a transmission priority and the different priority classes are considered.

5.   The intergeneration time for messages is a random variable with negative exponential distribution and a mean value μ, which is identical for all the stations. By varying the μ parameter it is possible to change the contribution of all system stations to the network load.

As is well known, the IEEE 802.4 standard specifies that a message can be transmitted only when the token is owned by the sending station. In the immediate acknowledgment (first) model the token owner sends a message and then waits for the acknowledgment frame from the remote stations. Then the time T_{st} required by a station to transmit a number k of messages once the token has been received and then to release the token itself is given by the following formula:

$$T_{st} = k(T_{mt} + T_{at} + 2T_d) + T_{tt} + T_d \qquad (1)$$

In equation (1) T_{mt} is the time period required by the station to transmit a single user message on the physical channel including the packet header and trailing bits, while T_{at} is the time needed to send the acknowledgment frame on the same channel. In addition T_{tt} is the time spent in passing the token to the subsequent station in the logical ring; in our model T_{tt} includes the *response window(s)* used to allow the dynamic insertion of new stations in the logical ring. Finally the term T_d takes into account the whole channel delay due to the transmission medium, the interface circuits and the head-end remodulator. Usually T_d is not negligible when the transmission frequencies suggested by the IEEE 802.4 standard are used. In the delayed acknowledgment (second) model a different policy is adopted for frame transmissions by the owner of the token. In fact, in this case, the station first sends the acknowledgment messages for all the frames received from the other nodes since the last holding of the token, then one or more user messages can be transmitted depending on the values of the token rotation time (TRT) and of the target token rotation time (TTRT) parameters [8]. The station transmission time when k acknowledgments and h user messages are sent can be computed as follows:

$$T_{st} = hT_{mt} + kT_{at} + T_{tt} + T_d \qquad (2)$$

It is worth noting that when $h = k$ the channel delay term (t_d) affects the value of T_{st} less in (2) than in (1). This is so because in the immediate acknowledgment model a response frame must be received before a new message can be transmitted. Instead, the delayed acknowledgment system allows a station to start the transmission of the next message immediately after the current frame has been sent. Messages can then be "pipelined" on the channel and the propagation delay is considered only when the pipe must be emptied (that is, when the token is passed to the next station).

3.2. Simulation Results

The results presented in this section were obtained assuming that the communication channel does not introduce errors so that every frame transmitted is always correctly received. This assumption allows us to perform a pure performance analysis, since it does not take account of the effect of the recovery mechanisms, and it makes the results obtained consistent with the assumptions found in several performance analyses of other networks presented in the literature, such as those presented in Pimentel, Damen, Killat, and Strecher, and Sachs, Kan, and Silvester [10, 11, 12].

We have selected the normalized response time experienced by messages in the network in order to study the behavior of the Type 3 LLC. It is worth noting, however, that in this case the definition of transmission delay must also take into account the time required to transmit the acknowledgment frame. Thus, in our system the actual data frames are considered separately from acknowledgment frames. The performance index used is given by the following formula:

$$\tau_n = E\left[\frac{t_{ack,i} - t_{gen,i}}{t_i}\right] \tag{3}$$

where:

τ_n is the normalized response time

$t_{gen,i}$ is the generation time for the ith data frame

$t_{ack,i}$ is the instant when the reception of the positive acknowledgment for the ith frame is completed by the station that originated the ith data frame

t_i is the time required to transmit the ith data frame.

By using a normalized index it is possible to obtain results that are independent of the physical channel bit rate. It is worth pointing out that a lower bound to the numerator in (3) is given by the addition of the time required to transmit a frame plus the time needed to transmit the acknowledge and possible propagation delays, as in (3) the transmission of a frame is not completed until the corresponding positive acknowledgment frame is correctly received; it turns out that the normalized response time is lower bounded by a value greater than 1, and this value increases with the ratio between the lengths of the acknowledge and data frames. The normalized response time is also the normalized average turnaround time observed by the user of the data-link layer selecting the type 3 service for the transmission of a packet (except for the time taken to cross the LLC sublayer, which is strongly implementation dependent).

In addition to the error-free assumption described above, the following properties have been introduced to obtain the results presented in this section:

1. the probability of addressing with data frames the different stations in the network is evenly distributed among all the possible destinations.
2. in the case of immediate acknowledgment, the station addressed by a data frame sends its positive acknowledgment frame after a time of 3 octets from the end of the received data frame.
3. all the data frames have the same length (for each simulated experiment), while the acknowledge frames have the same length as the token frame.

Assumption 3 has been introduced in order to study the effect of the data frame length on network performance, since this requires only long or short frames to be generated during the same simulated experiment.

All the results, which were obtained for different traffic conditions, are expressed by the following parameter:

$$O_I = \frac{L_f}{\mu T_r} \tag{4}$$

where

O_I is the offered load
μ is the mean value of the inter-generation time for messages
L_f is the frame length
T_r is the transmission rate in bits per second.

The dividend in (4) also includes all the control bits of the data frames, while all the bits of acknowledgment and token frames are not counted; with this definition, the offered load is proportional (with a coefficient depending on the message length) to the ratio between the number of useful bits generated and the nominal bandwidth.

Figure 8.9 shows the values of the normalized response time obtained for both immediate acknowledgment (IA) and delayed acknowledgment (DA); two sets of curves are reported: curves (a) concern the transmission of short (200-byte) frames, while curves (b) illustrate the system's behaviour when long (1024-byte) frames are considered.

The results in Figure 8.9 were obtained in the ideal case of a zero propagation delay on the channel. In this ideal case, IA outperforms DA in all traffic conditions, with a much improved performance in heavy traffic.

The hypothesis of no transmission delay on the channel is not realistic, since the physical medium and the modulators as well as head-end, in the case of broadband transmission, introduce delays that are not negligible with respect to the time required to transmit a bit. Therefore, it is useful to study the influence of the transmission delay on the performances of both forms of the acknowledged datagram service.

The results of such investigations are reported in Figure 8.10, where the transmission delay is always equivalent to 30 μs (300 bits for a 10 MHz network or 150 bits for a 5 Mb/s carrier band), and in Figure 8.11, where the transmission delay is always equivalent to 60 μs.

When the transmission delay introduced by the channel itself increases, both IA and DA show poorer performance; however, the performance falloff is larger for IA than for DA, and this effect is more relevant for heavy loads. As shown in Figure 8.10 and Figure 8.11, the introduction of increasing transmission delays leads the performance curves of IA and DA to intersect and this does not appear in the ideal case of Figure 8.9. Furthermore, this intersection moves toward light loads as the transmission delay is increased.

The reason for this behavior is intrinsic in the two types of MAC protocols. In the IA case, an acknowledgment after the transmission of each data frame is required; this means that the activities are suspended for the period necessary for

234

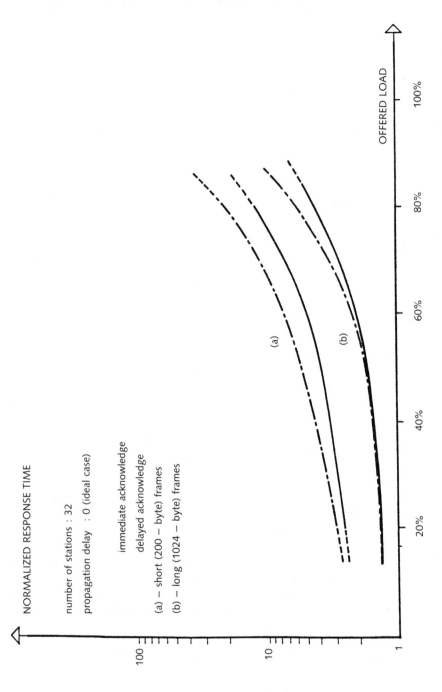

NORMALIZED RESPONSE TIME

number of stations : 32
propagation delay : 0 (ideal case)

immediate acknowledge
delayed acknowledge

(a) – short (200 – byte) frames
(b) – long (1024 – byte) frames

OFFERED LOAD

FIGURE 8.9. Performance of the immediate acknowledge and delayed acknowledge systems with no propagation delay (ideal case).

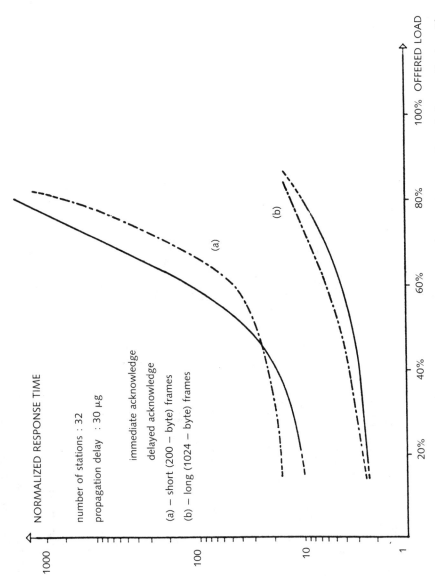

NORMALIZED RESPONSE TIME

number of stations : 32
propagation delay : 30 μg

immediate acknowledge
delayed acknowledge

(a) — short (200 — byte) frames
(b) — long (1024 — byte) frames

FIGURE 8.10. Performance of the immediate acknowledge and delayed acknowledge methods with propagation delay equal to 30 μs.

235

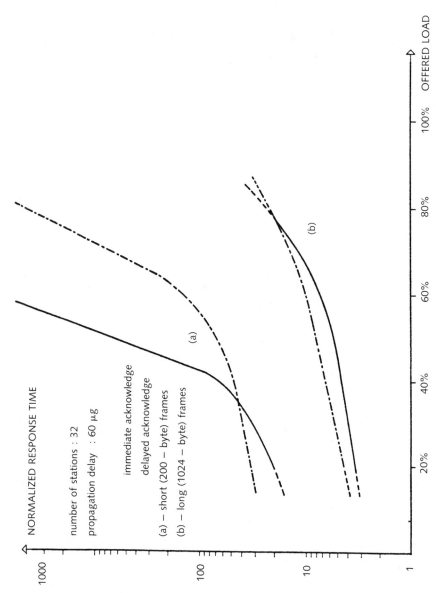

NORMALIZED RESPONSE TIME

number of stations : 32
propagation delay : 60 μg

immediate acknowledge
delayed acknowledge

(a) — short (200 — byte) frames
(b) — long (1024 — byte) frames

(a)

(b)

20%   40%   60%   80%   100%   OFFERED LOAD

1000

100

10

1

FIGURE 8.11. **Performance of the immediate acknowledge and delayed acknowledge methods with propagation delay equal to 60 μs.**

the data frame to propagate to the destination, and for the acknowledgment to return. This idle time is repeated for each data frame transmitted by a station during the period it holds the token.

Conversely in the DA case, a station holding the token can transmit, without interframe gaps due to propagation, several data and acknowledgment frames and this can be done until the queues of frames waiting for transmission are exhausted or the preset token holding time is terminated. It should be noted that in heavy load conditions there is a high probability of multiple-frame transmission, since it is likely that a station receiving the token needs to transmit one or more acknowledgment and data frames. This occurs even though the token rotates with a period close to the target token rotation time for it is often possible to have enough time to pack an acknowledgment (short frame) and a data frame in the same transmission. The overall result is a better utilization of the channel bandwidth and a shorter average turn around time for the acknowledged datagram service, when short frames are used (see curves (a) in Figures 8.9–8.11).

Further support for the explanation of the behavior of IA and DA techniques for different values of the channel delay, is given by the results which were obtained for long data frames, shown by curves (b) in Figure 8.9–8.11. In this case, the same value of the offered load is obtained with longer and fewer messages, therefore the number of idle periods spent waiting for the acknowledgment, in the IA case, is lower than for short frames. It turns out that the effect of performance falloff with respect to DA due to an increase in the channel delay, is lowered, and the intersection points for the two curves disappear or are found for higher values of the offered load, with respect to the results shown for short frames.

3.3.   Message Priorities

A relevant feature of the medium-access mechanisms developed for token bus networks is the possibility of using priority levels in order to obtain shorter waiting times for urgent messages.

The type of priority mechanism considered in the following is that of the IEEE 802.4 standard [8]. This mechanism has four levels of priority (6, 4, 2, and 0, in decreasing order of priority), and each station in the network is seen as consisting of four logical stations, each one associated with a priority class. Whenever the token reaches a station, the token rotation time of the last tour is measured and subtracted from the target token rotation time (TTRT); if the difference is positive, then the station can hold the token for transmitting at least one highest priority message.

Once pending messages with the highest priority have been exhausted before the expiration of the holding time, a virtual token is passed to the logical station with the next lower priority; then the token rotation time is recomputed (to take account of the time consumed by the higher priority messages transmitted) and subtracted from the time threshold associated with the specific priority level. If the result is positive, then it is possible to transmit messages of that priority class until the threshold is passed, otherwise no transmission is possible and the virtual token is passed to the next lower priority logical station. This process is repeated within a station until the lowest priority station is reached; then the real token is passed to the next station in the token logical ring.

Two important considerations arise from the brief description of the priority mechanism given above:

- each priority class has its own time threshold, whose value influences the behavior of priority classes and should be constrained by some relations (for example, the lower the priority class, the lower the threshold value)
- the mechanism does not guarantee a real enforcement of priority privileges in some situations.

The second consideration can be illustrated by means of an example. Let us consider two consecutive stations in the token logical ring, i and $i - 1$, and assume that station i has only a class 2 message waiting for transmission, while station $i - 1$ has only one message pending of class 4. Assume that in the previous tour, no message has been transmitted by the two stations, so that the time of the last token visit differs in the two stations only by a small amount δ. If the token reaches i just in time to allow the transmission of the class 2 message, the station $i - 1$ will measure a value of the token rotation time greater than that of i, and the difference is T (a δ time is always required to pass the token). Thus, as the measured rotation time is increased, its value can be greater than the threshold for class 4, and the pending message cannot be transmitted because it has been blocked by the transmission of a lower priority one.

The values assigned to the time thresholds for the different priority classes are critical for the correct behavior of the network; this point is illustrated by the examples of bad threshold selection shown in Tables 8.2 and 8.3. In the former, though the threshold values are not equal, there is no distinction between the values of the normalized response time for the different priority classes (except for the highest one) for both IA and DA, because the average token rotation time is far lower than the smallest threshold, so that it is always possible to transmit the pending messages, regardless of the priority class they belong to; it turns out that there is still a priority ordering within each station, but all the messages are transmitted when the token is held by the station. Conversely, the results in Table 8.3 are obtained by using very tight thresholds; the effect is that no message of classes other than the highest priority one are transmitted; in this case the average token rotation time is longer than the greatest threshold (but lower than TTRT); hence, on average, there is no chance, even for the messages of the next to

TABLE 8.2. Simulation parameters and results for a network with 32 stations, *offered load* = 60% and *propagation delay* = 30 μs

Model Type	Priority Class	Generated Messages	Transmitted Messages	Threshold Value	t_{ave} Value	Normalized Response Time
	6	12934	12930	9.984		47.34
Imm.	4	13096	13080	30.00	14.595	100.04
Ack.	2	13321	13307	25.00		100.54
	0	13229	13211	20.00		100.52
	6	12934	12931	8.064		38.42
Del.	4	13096	13087	30.00	6.218	56.72
Ack.	2	13321	13316	25.00		54.68
	0	13229	13223	20.00		56.13

highest priority class, to be transmitted. Only the class 6 messages are sent because they use the TTRT as a threshold value.

The choice of the threshold values, therefore, involves a tradeoff between differentiation among classes and the possibility of transmitting all the messages generated. The former effect increases when thresholds get tighter, the second when thresholds are not tighter than a given bound. We now try to give an estimation of such a lower bound for the class thresholds.

First, it should be noted that the lower bound is referred to as the threshold of class 0, because all the other threshold values are greater.

The average token rotation time (t_{ave}) can be expressed as follows:

$$t_{ave} = t_{0l} + \sum_{j=0}^{3} f_{2j} \qquad (5)$$

where f_{2j} is the time spent, on average, at each rotation to transmit messages of class 2_j and t_{0l} is the zero-load latency time of the logical ring. Though the

TABLE 8.3. Simulation parameters and results for a network with 32 stations, *offered load* = 60% and *propagation delay* = 30 μs

Model Type	Priority Class	Generated Messages	Transmitted Messages	Threshold Value	t_{ave} Value	Normalized Response Time
	6	12934	12934	9.984		10.25
Imm.	4	13096	1183	2.400	2.950	—
Ack.	2	13321	6	2.350		—
	0	13229	0	2.300		—
	6	12934	12934	8.064		16.61
Del.	4	13096	1376	2.400	2.756	—
Ack.	2	13321	3	2.350		—
	0	13229	0	2.300		—

priority mechanism does not strictly enforce the priority rules, it does when the average behavior is considered. In equation (5), the contributions due to the different classes are explicitly given by the terms of the summation; thus, assuming that, in order to transmit class 0 messages, it is necessary for the token rotation time to allow the transmission of all the messages in the other classes, it follows that the contribution to the average token rotation time given by t_{ol} and by all the classes from 2 to 6 bounds the threshold for class 0, according to the following formula:

$$t_{thr}^{(0)} \geq t_{ol} + \sum_{j=1}^{3} f_{2j} \qquad (6)$$

Note that the bound expressed by (6) is not guaranteed to be tight; nevertheless it is possible to derive similar bounds for the other classes by excluding other terms from the summation in (6).

The effect of varying threshold values for prioritized traffic was analyzed by direct simulation for both the immediate and delayed acknowledgment schemes. In particular experiments were conducted to study how many messages of a given class, which are generated by each station, can be transmitted when the associated threshold value becomes lower than the corresponding bound obtained from (6) or its modifications for classes other than 0.

Figure 8.12 shows the results obtained for the IA model, while Figure 8.13 refers to the DA case. For each priority class the fraction of messages generated in the network that was successfully transmitted is reported versus the signed difference between the threshold value selected for each class and the bound computed with an "a posteriori" analysis of the simulator behavior. This difference is normalized by the difference between the bound value and the minimum token rotation time, that is, the rotation time when no message is sent through the network. Figure 8.12 and Figure 8.13 show two sets of curves which have been obtained for different load conditions (*offered load* = 20% and *offered load* = 60%, respectively).

In both cases the behavior of the system is the same. The percentage of transmitted messages is nearly 100% when the threshold value is greater or equal to the bound value, while it decreases rapidly when the threshold becomes lower than the computed bound. Therefore, the simulation results are in good accordance with the bound obtained in (6) for class 0, and with the bounds for the other classes that can be obtained by modifying (6). It is worth noting that (6) is similar to the bounds obtained from the models presented in Gorur and Weaver [13] and in Jayasumana and Fisher [14] for the 802.4 with LLC Type 1.

Simulation results show that the formulae in Gorur and Weaver [13] and Jayasumana and Fisher [14] can also be used in our case to determine a priori the threshold values, provided that the following changes are introduced. In the IA case the frame transmission time must include the overhead introduced by the transmission of the acknowledgement message and twice the propagation delay

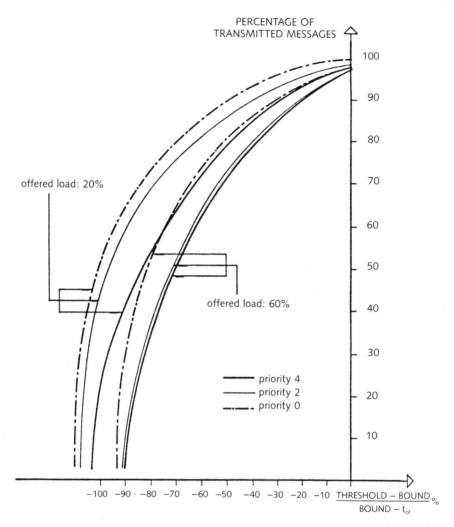

FIGURE 8.12. Percentage of transmitted messages versus the difference between the threshold value and the observed TRT in IA mode.

value. In the DA case the results obtained are in line with the model in Gorur and Weaver [13] when the acknowledgment transmission time is added to the frame transmission time.

It can be added that (6) can be usefully adopted in network management activities as it can be used, applied to finite time intervals, to detect when anomalous performance conditions connected with the priority mechanism are likely to arise. Furthermore, (6) indicates which parameters should be monitored in order to implement performance management in token bus networks.

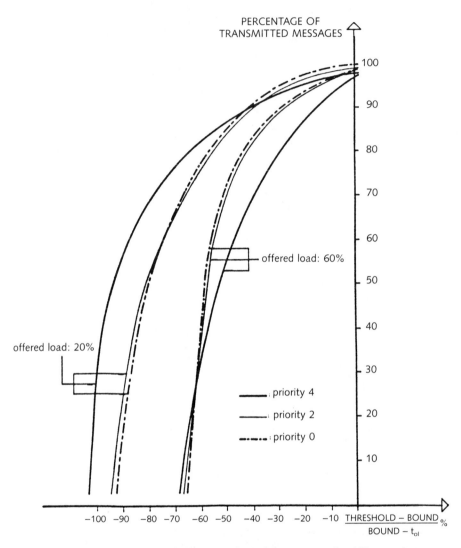

FIGURE 8.13. Percentage of transmitted messages versus the difference between the threshold value and the observed TRT in DA mode.

4.   IEEE 802.5 PERFORMANCE ANALYSIS

4.1.   The Model

The model of the system is based on the same assumptions introduced in the previous section, but the N identical stations are now connected by a single communication ring based on the IEEE 802.5 standard specifications [9].

Furthermore, since the 802.5 standard allows messages belonging to eight priority classes to be exchanged, the acknowledgment frames can be assigned, in principle, to different priority classes, too.

Also, in this case the response time for a message is assumed to be equal to the time interval between the message generation time (i.e., the moment at which the message is passed to the MAC by the LLC entity) and the reception by the MAC sublayer of the last bit in the corresponding acknowledge from the remote stations.

4.2. Simulation Results

This section presents the simulation result obtained for a token-ring network where all the messages are transmitted with equal (highest) priority. Each transmitted message is assumed to be correctly received (i.e., the communication channel does not introduce errors). Also, each station which receives a message returns a positive acknowledgment to the sender as soon as possible. Furthermore, it is also assumed that the acknowledgment messages do not experience any transmission error.

The performance analysis has been carried out ignoring the effects of the recovery mechanisms introduced in the various protocol layers. With this hypothesis, the results obtained are consistent with the typical assumptions on which several network performance analyses that have been presented in the literature, are based; such as those presented in Pimentel [10], Damen, Killat and Strecher [11], Sachs, Kan and Silvester [12], Karvelas and Leon-Garcia [15] and Chauhan and Sethi [16].

Figure 8.14 shows the values of the normalized response time obtained in a network consisting of 40 stations; two curves are reported: curve (a) refers to the transmission of short (100-byte) frames, while curve (b) illustrates the system's behaviour when long (1024-byte) frames are considered.

For any given value of O_l the system's performance is better for long frames than for short frames. This is so since in curve (b) the same offered load is obtained with longer and fewer messages than in curve (a); therefore the number of idle periods, spent waiting for the acknowledgment, is lower for long frames than for short frames.

Another parameter that can affect the performance of a real system is the maximum token holding time (THT). Since the THT must be selected by all the station when the network is started, it is interesting to predict, to some extent, how the system will behave for different values of THT, so that the best choice can be made according to the application needs.

The curves in Figure 8.15 show the normalized transmission delay versus the offered load time for 100-byte messages and for different values of THT. It is worth noting that when THT is such that a station is forced to pass the token after the transmission of a single message, the maximum transmission delay is experi-

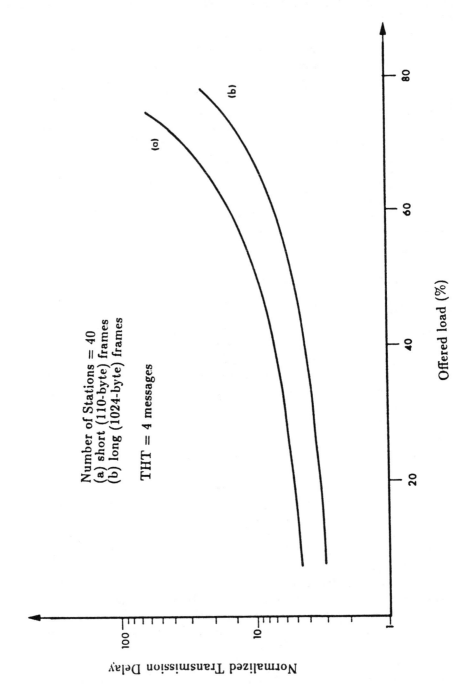

Number of Stations = 40
(a) short (110-byte) frames
(b) long (1024-byte) frames

THT = 4 messages

(a)

(b)

Normalized Transmission Delay

Offered load (%)

FIGURE 8.14. Performance of Type 3 LLC for a token ring network with a single priority class of messages.

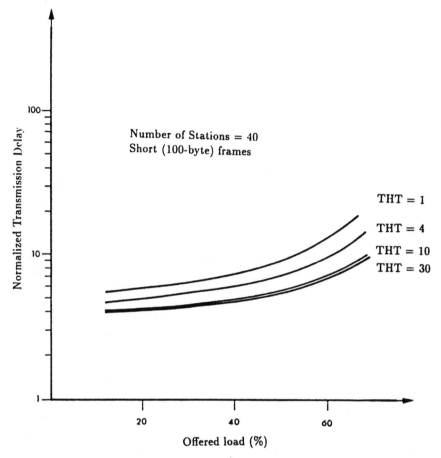

FIGURE 8.15.   Effect of the token holding time on system performance.

enced, while the transmission delay is reduced when THT is increased. However, when THT is long enough to allow the transmission of 10–12 messages
without passing the token, the system's performance is no longer enhanced by
increasing the value of the token holding time.

Thus a practical upper bound for THT is established; as using THT values
beyond this bound does not reduce the mean system response time, but, instead,
the maximum token rotation time is increased. Such behavior may not be accept-
ed in industrial environments where real-time is a crucial issue.

4.3.   Message Priorities

Unlike other token-based access methods, such as IEEE 802.4 [8] and FDDI
[17], IEEE 802.5 uses explicit reservation fields in the frame; each station, with

a message pending at a priority level higher than the reservation field received, increases the reservation code in order to establish the maximum priority of the pending messages.

All the tokens transmitted on the ring have a priority level encoded in them, so that only those stations having frames pending at priority levels equal to or higher than the token priority are allowed to transmit.

The above discussion is just a brief description of the actual priority mechanisms used by IEEE 802.5; other operations are required by the standard and all these mechanisms have been implemented in simulation programs.

The results for the token ring with priorities have been obtained under the same general assumptions as those used earlier in this chapter. In particular, the acknowledgment of the receipt of a data frame, required by the DL-DATA-ACK service, is performed by sending a short response frame, while the DL-DATA and DL-REPLY are not considered in this chapter.

The introduction of priorities leads to the problem of assigning a priority for acknowledge frames, too. A possible criterion considers all the acknowledge frames as urgent, thus it assigns them the highest possible priority reserved to nonsynchronous traffic. An alternative is to assign the acknowledge frames with the same priority as the corresponding request frame.

The above alternative is adopted in our model, because the acknowledge frames at lower priority levels do not cause an increase in traffic at higher priority levels. In this way, heavy load conditions at low priority have only a slight effect (or no effect at all) on the delivery time of higher priority frames. This separation would not be maintained, if high priority levels were assigned to all the acknowledge frames.

The number of priority levels allowed in the IEEE 802.5 standard is 8; however, in our simulations, we have introduced only 4 due to the fact that 4 priority levels are used in other standards for industrial automation, such as in IEEE 802.4 [8] which is used in MAP [1] and Proway [18]. With this choice, the results are comparable with those obtainable for other standards, and the number of priority levels is always sufficient for industrial control applications.

The normalized delay for different load conditions and for the 4 priority classes is shown in Figure 8.16, for frames with 100 bytes of data, and in Figure 8.17, for frames with 1024 data bytes. Comparing the effect of prioritization with that obtained for IEEE 802.5 without acknowledgment in Peden and Weaver [19] it can be noted that, while in Peden and Weaver the maximum difference between the delay of the highest and lowest priorities is a factor of only 2.5 for 40 stations and short frames, this difference in our case is increased up to more than one order of magnitude.

This result is not unexpected, as the performance index also takes into account the delay due to control frames (i.e., acknowledgments); this confirms the feeling that the effect of prioritization at the medium access level has greater consequences at higher protocol layers, because the delay, from the point of view of a

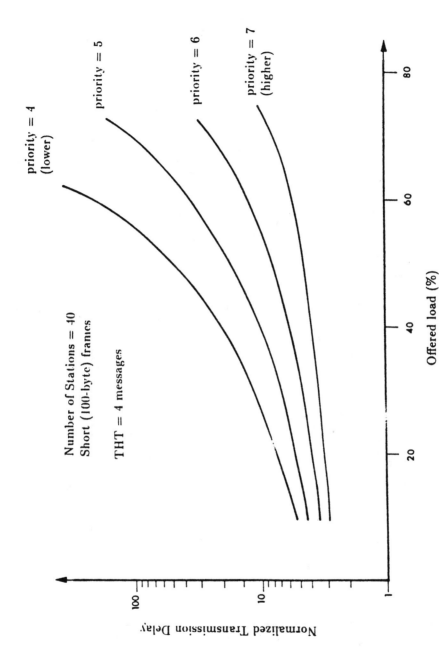

FIGURE 8.16.   Performance of a token ring network using four priority classes for short frame messages.

247

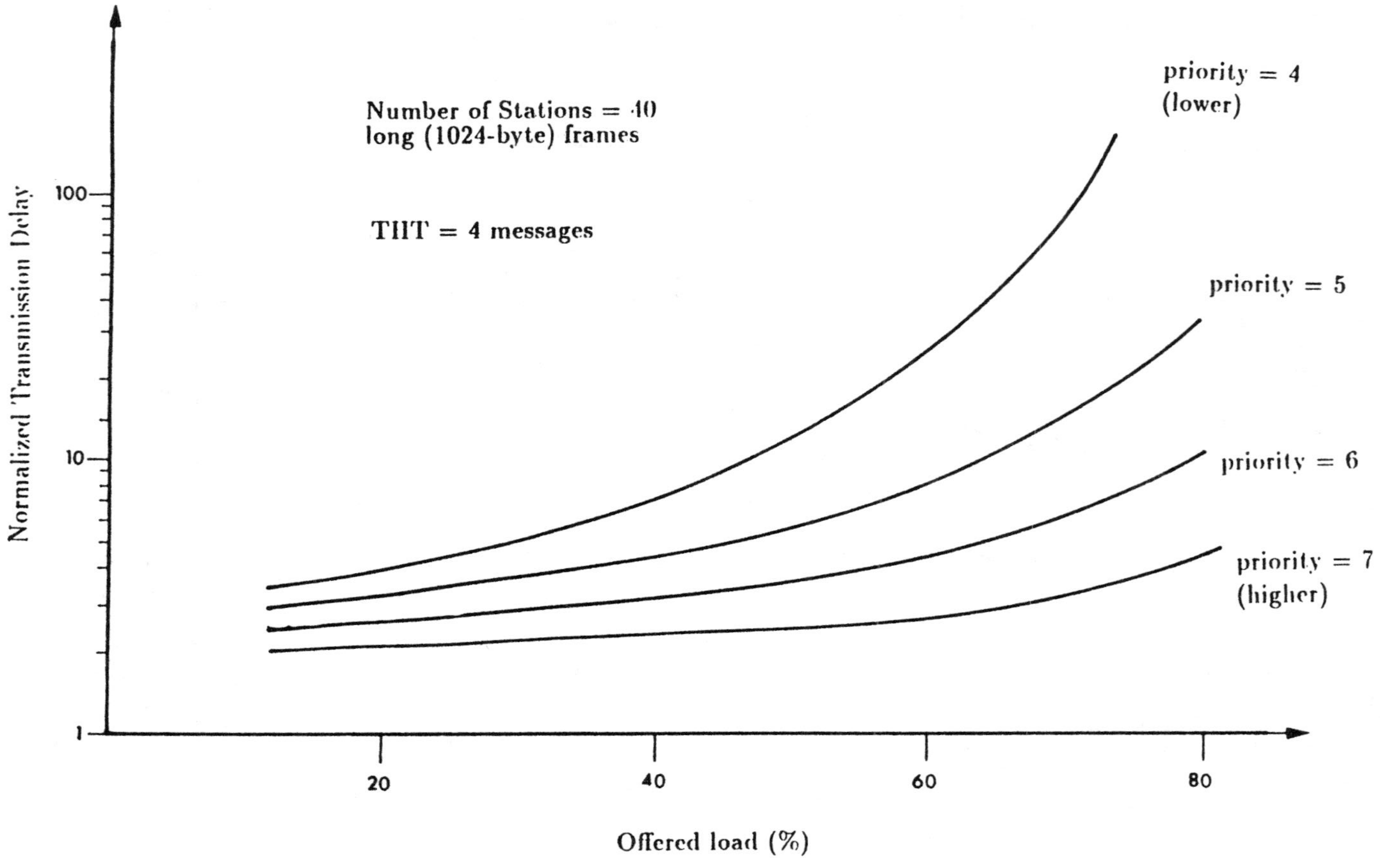
Number of Stations = 40
long (1024-byte) frames
THT = 4 messages
priority = 4
(lower)
priority = 5
priority = 6
priority = 7
(higher)
Normalized Transmission Delay
100
10
1
20
40
60
80
Offered load (%)

FIGURE 8.17. Performance of a token ring network using four priority classes for long messages.

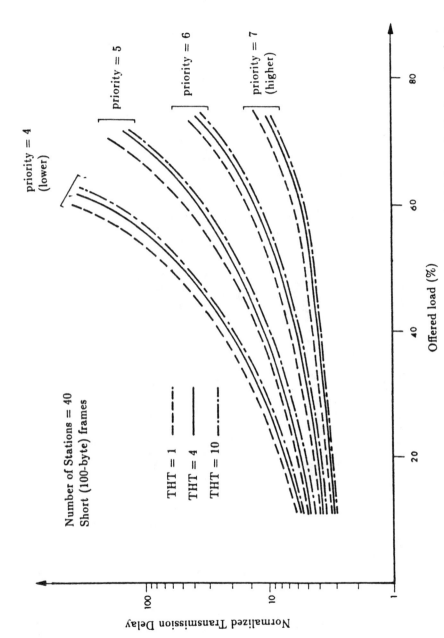

FIGURE 8.18. Effect of the token holding time on the performance of a network with four priority classes for short messages.

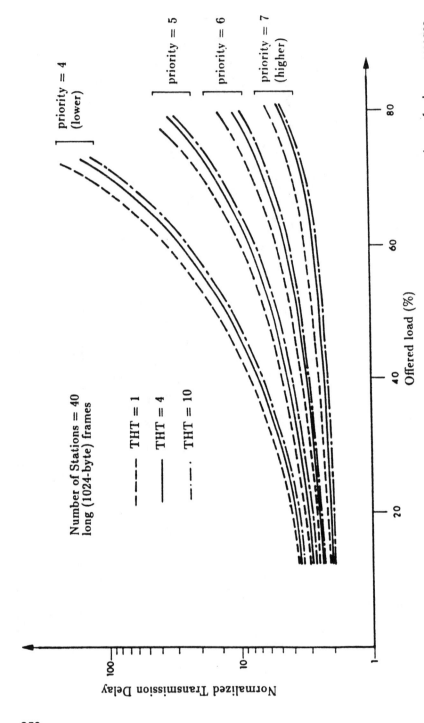

FIGURE 8.19. Effect of the token holding time on the performance of a network using four priority classes for long messages.

user of an upper protocol layer, includes the time for the transmission of the data as well as for several low-level control frames.

The curves in Figure 8.16 and Figure 8.17 show that now the 4 curves representing the different classes are spread around the values of the curve obtained for the system without priorities.

The curves in Figure 8.18 and Figure 8.19 show the results obtained by varying the value of the token holding time. The results measuring the effect of such a parameter are similar to those obtained when the simulation is carried out for a system with no priority classes; increasing the value of THT leads to better average performances for any value of THT up to approximately 10–12 messages, while this effect disappears for larger values.

REFERENCES

1. MAP Task Force, *Manufacturing Automation Protocol, Version 3.0, Implementation Release.* Warren, MI: General Motors Technical Center, 1987.
2. ISO/TC97/SC16/N.7498, *Open Systems Interconnection Architecture*, 1981.
3. ISO, *Information Processing Systems—Open Systems Interconnection—Basic Reference Model,* International Standard No. 7498, 1984.
4. ANSI/IEEE Std. 802.2, *Logical Link Control*, 1985.
5. ISO DIS 8802/2, *Logical Link Control*, 1986.
6. ISO /TC97/SC6/WG1, *Draft Proposed Addendum to ISO DIS 8802/2 Logical Link Control—Acknowledged Connectionless Service*, 14th Draft, June 1986.
7. ANSI/IEEE Std. 802.3, *CSMA/CD Access Method*, 1985.
8. ANSI/IEEE Std. 802.4, *Token-Passing Bus Access Method*, 1985.
9. ANSI/IEEE Std. 802.5, *Token Ring Access Method and Physical Layer Specifications*, 1985.
10. J.R. Pimentel, "Performance Evaluation of MAP Networks," *Proceedings of IECON '85*, 1985, pp. 629–634.
11. N. Damen, U. Killat and R. Strecher, "Performance Analysis of Token-Bus and CSMA/CD Protocols Derived from FORCASD Simulation Runs," *Performance of Computer Communication Systems.* Amsterdam: North-Holland, 1984, pp. 79–94.
12. S.R. Sachs, K. Kan and J.A. Silvester, "Performance Analysis of a Token Bus Protocol and Comparison with Other LAN Protocols," *Proceedings of the 10th IEEE Conference on Local Computer Networks*, 1985.
13. M. Gorur and A.C. Weaver, "Setting Target Rotation Times in an 802.4 Network," *Proceedings of the Workshop on Factory Communications*, NBS Internal Report 87-3516, March 1987, pp. 199–220.
14. A.P. Jayasumana and P.D. Fisher, "Performance Modeling of IEEE 802.4 Token Bus," *Proceedings of the Workshop on Factory Communications*, NBS Internal Report 87-3516, 1987, pp. 221–245.
15. D. Karvelas and A. Leon-Garcia, "Performance of Integrated Packet Voice/Data Token-Passing Rings," *IEEE Journal on Selected Areas in Communication*, Vol. SAC-4, No. 6, 1986, pp. 823–832.

16. V. Chauhan and A.S. Sethi, "Performance Studies of Token Based Local Area Networks," *Proceedings of the 10th IEEE Conference on Local Area Networks*, 1985, pp. 100–107.

17. ANSI, *FDDI Token Ring Media Access Control*, Draft X3T9.5/83-16, Rev. 8, March 1985.

18. Proway SC65/WG6, *Specifications for Highway Unit Protocol*, IEC, Vol. 1, Part 3, 1985.

19. J.H. Peden and A.C. Weaver, "Are Priorities Useful in an 802.5 Token Ring?," *Proceedings of the Workshop on Factory Communications*, NBX Int. Rep. 87-3516, March 1987, pp. 253–270.

9

A Programming Methodology
for Robotic Arc Welding*

Kristinn Andersen
George E. Cook
Saleh Zein-Sabattou
Robert Joel Barnett
Department of Electrical and Computer Engineering
Vanderbilt University
Nashville, TN

Kenneth R. Fernandez
National Aeronautics and Space Administration
Marshall Space Flight Center
MSFC, AL

1. OVERVIEW

This article presents algorithmic approaches for the programming of robots with multiple and redundant axes of motion. The systems of particular interest are simultaneously moving welding robots and workpiece positioners. The following features are provided by the presented algorithms:

1. Torch Position Control,
2. Downhand Welding Control,
3. Wire Guide Orientation Control, and
4. Controllable/Constant Welding Speed.

Both the theoretical framework and simulation results, demonstrating the programming algorithms, are given in the article.

The practical problems reduced or eliminated by the proposed algorithm are briefly summarized in the following. Each teach point along any robotic welding

---

* This research has been partially supported by NASA Contract No. NAS8-37629.

path has to be programmed with certain constraints satisfied. The workpiece has to be appropriately oriented, the torch tip has to be appropriately positioned and oriented, the wire-feed mechanism should be oriented in a certain way, and, furthermore, welding speed is to be controllable at all times. The first three requirements can be met with considerable patience on the programmer's behalf using a robot and a positioner, manually guided with a teach pendant. Although viable, this approach is tedious and prone to programming errors. In an integrated CAD/CAM system, on the other hand, this step may not be easily achievable without an algorithm, such as the one presented here. Most welding systems have more than six degrees of freedom and the problem of how to coordinate the redundant joints and simultaneously satisfy the welding constraints has to be resolved. The fourth requirement, programmability for constant or controllable welding speed, using arbitrary teach point spacing, is neither provided for by the teach-pendant approach nor the unmodified CAD/CAM approach.

A programmer using the proposed algorithm can save substantial time and ensure accuracy by applying the algorithm as follows. In the case of programming the weld path by guiding the robot with a teach pendant, the programmer can orient the workpiece by aligning the positioner joints into any convenient position at the beginning of the programming sequence, as well as at any time throughout the programming process. The torch is guided to the first teach point on the workpiece where the programmer only needs to ensure adequate proximity to the desired teach-point location and the desired torch orientation with respect to the workpiece. Wire-feed orientation and workpiece orientation are not relevant here. This is repeated for all teach points. In the case of a CAD/CAM system, the coordinates and desired torch orientations at the selected teach points can be extracted from the workpiece data, from which all information for the algorithm can be derived. In either case, the desired welding speed, or speeds, are also specified. From this information the proposed welding algorithm calculates the desired positioner and robot joint angles, and robot speeds required to make the weld with the desired welding constraints. In short, a process which was, at best, a series of tedious trial-and error operations is turned into an efficient one-pass task using the presented algorithms.

2. INTRODUCTION

Arc welding is among the most common metal joining methods used in modern manufacturing facilities. During the past decade, robots have increasingly taken over many welding tasks that used to be carried out by human welders. The use of robots for repetitive welding operations and applications that call for high precision has, in many cases, been justified by their effectiveness as compared to human welders.

While industrial robots are advantageous in that they can carry out tasks with

high speed and precision for extended periods of time, they have shortcomings of
their own as well. Programming a robot for a specific task is frequently time
consuming and usually it is necessary to test and modify the programmed sequence a number of times until the desired motions are achieved. This chapter
considers the simultaneous coordinated control of robotic mechanisms and positioners to satisfy geometrical constraints arising in robotic arc welding. The
various arc welding processes, such as Gas Tungsten Arc Welding (GTAW) and
Variable Polarity Plasma Arc Welding (VPPAW), demand the welded workpiece
and the welding torch to be appropriately positioned and oriented at all times
during welding to yield satisfactory welds. Additionally, the speed of the torch
tip, with respect to the welded joint, has to be fully controllable.

In a typical robotic welding assembly the workpiece is held by a *part positioner,* which may be viewed as an N-Degrees-Of-Freedom (N-DOF) robotic
mechanism itself. The *robot* (*manipulator*) holds the welding torch including the
electrode, shielding gas nozzle, wire feeder which supplies the molten weld pool
with reinforcement material, and other associated hardware as applicable. To
obtain a satisfactory weld, the torch and the workpiece have to be manipulated in
certain manners depending on the applied welding process. The first requirement
is that the torch tracks the joint to be welded within acceptable tolerances.
Secondly, the orientation of both the workpiece and the torch has to be appropriate for the welding process. In most cases, the workpiece should be oriented so
that the surface tangent to the workpiece, where the molten weld pool is located,
is always horizontal. In some welding applications, however, the welded spot is
preferred to be on a vertical, rather than horizontal, surface at all times, and,
furthermore, the workpiece may have to be rotated so that the molten spot moves
vertically up during the entire welding pass. These workpiece orientation requirements are usually accomplished through continuous part rotation with the
positioner. Regarding the torch manipulator, there are a number of requirements,
or constraints, which have to be met as well. Usually the filler wire is fed from
the front of the arc plasma as it moves along the welded joint. The angles
between the arc-plasma flow, the workpiece surface, and the instantaneous direction of the torch travel may have to maintained at specific values of orientation
during the entire weld pass. Finally, the speed of the torch tip, with respect to the
workpiece surface, has to be completely controlled. Typically, the speed is kept
constant at a value selected by the operator. For applications where the positioner
is stationary during the entire welding pass, this is a trivial task. In cases requiring simultaneous, coordinated movements of the positioner and the manipulator,
however, speed control becomes more complex. The simultaneous motion of the
positioner and the manipulator constitute complications for programming the
teach points for the positioner and the manipulator as well. Solutions to these
dilemmas are presented in this section.

Before embracing upon the problems at hand and the solution approaches, a
brief overview of some basics of robotic manipulator control is in order. General,

and more detailed, treatments on robotic manipulators can be found in Paul [1] and Critchlow [2]. Specific treatment of robotic welding is given in Lane [3]. The notations and terminology adhered to here are primarily derived from Paul's textbook.

The position of an object in three-dimensional space can be specified by reference to a fixed Cartesian coordinate system, the *world coordinate system*. A point, or a vector u, in the space is specified here in terms of its x, y, and z coordinates, and a scaling parameter, w. In matrix form, u is denoted as $[x,y,z,w]^T$. The actual position of u in the space is obtained by dividing the coordinate parameters by the scaling parameter. Therefore, a specific vector u can be represented as a 4×1 vector in an infinite number of ways:

$$u = [2,3,5,1]^T = [4,6,10,2]^T = [1, 1.5, 2.5, 0.5]^T \tag{1}$$

A scaling parameter of $w = 1$ will be used in all simulations and examples here, unless otherwise indicated. A *translation* of the vector u by the vector (a,b,c) is obtained by premultiplying u by a 4×4 transformation matrix:

$$u_{tr} = \begin{bmatrix} 1 & 0 & 0 & a \\ 0 & 1 & 0 & b \\ 0 & 0 & 1 & c \\ 0 & 0 & 0 & 1 \end{bmatrix} \cdot u \tag{2}$$

where u_{tr} is the resulting translated vector. *Rotation* of a vector about any of the coordinate axes is accomplished by a multiplication by a 4×4 matrix as well. For example, rotation of u by an arbitrary angle θ about the z-axis is accomplished by the following multiplication:

$$u_{rot} = \begin{bmatrix} \cos\theta & -\sin\theta & 0 & 0 \\ \sin\theta & \cos\theta & 0 & 0 \\ 0 & 0 & 1 & 0 \\ 0 & 0 & 0 & 1 \end{bmatrix} \cdot u \tag{3}$$

where u_{rot} is the rotated vector. Similar rotation matrices are readily found for rotations about the x- and y-axes. Any vector is specified in terms of its orientation as well as position. This is easily visualized by considering a local coordinate system, or frame, assigned to each vector and the origin of the coordinate system is placed at the coordinates of the vector. When a vector is translated from one position to another its coordinate system is translated, but its orientation remains intact. During rotation, however, the orientation changes. In the following, the terms *position* and *orientation* will be used in their conventional sense, while the term *location* may be used to refer to the combined state of position and orientation. If a vector u is translated and/or rotated successively by matrices A_1, $A_2, \ldots, A_n$, in that order, the resulting vector is obtained by the following premultiplications:

$$u_{new} = A_n A_{n-1} \ldots A_1 \cdot u \qquad (4)$$

where each translation or rotation is performed with respect to the same, fixed, world coordinate system. The identical net transformation can also be carried out by applying the n transformations in reverse order, but then the successive transformations must be executed with respect to the transformed coordinate systems at each step, rather than the world coordinate system. Assume that the vector u is assigned its own coordinate system, which is oriented identically to the fixed-world coordinate system. The first transformation is carried out with respect to the coordinate system of u. This coordinate system is transformed with u to a new position and/or orientation. The second transformation is done with respect to this transformed coordinate system. This is replaced in the same manner until all transformations have been carried out. Therefore, the total transformation is:

$$u_{new} = B_1 B_2 \ldots B_n \cdot u \qquad (5)$$

where $B_i = A_i$ for all i, but while the A transforms are performed with respect to the world coordinate system, the B transforms are performed with respect to the transformed coordinate systems. The transformation sequence of Equation (4) is referred to as *premultiplication*, while Equation (5) describes the *postmultiplication* sequence.

A robotic manipulator consists of a number of *links* connected by an equal number of actuated *joints*. A manipulator using N links (and joints) is referred to as an *N-degrees-of-freedom* (*N-DOF*) manipulator. The base of the manipulator can be regarded as link 0, but it is not considered as one of the actual manipulator links. Joint 1 connects link 1 of the manipulator to the base (link 0) and similarly any joint n connects link n to link $n - 1$. As far as the mathematics of the manipulator kinematics are concerned, each link is completely characterized by its *length*, a, and its *twist*, α, as illustrated in Figure 9.1. A joint may be either revolute or prismatic. A revolute joint is entirely specified in terms of the variable θ and the fixed parameter d, as shown in Figure 9.2. For a prismatic joint: d is the variable parameter, θ is fixed, and α is unspecified for the following link, as it has no meaning in that context. These parameters are further explained by Denavit and Hartenberg [4]. Using the general transformation matrix notation described earlier, one can specify the location and orientation of a coordinate system of any given link of a manipulator with respect to the previous link. If the

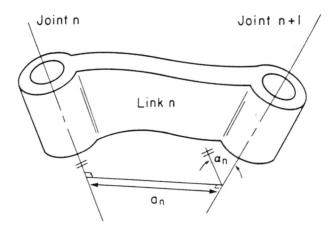

FIGURE 9.1.   Denavit–Hartenberg Parameter Definitions for a Robotic Link

links are connected with revolute joints the transformation relating link n to link $n - 1$ is:

$$A_{n,rev} = \begin{bmatrix} \cos\theta & -\sin\theta\,\cos\alpha & \sin\theta\,\sin\alpha & a\,\cos\theta \\ \sin\theta & \cos\theta\,\cos\alpha & -\cos\theta\,\sin\alpha & a\,\sin\theta \\ 0 & \sin\alpha & \cos\alpha & d \\ 0 & 0 & 0 & 1 \end{bmatrix} \quad (6)$$

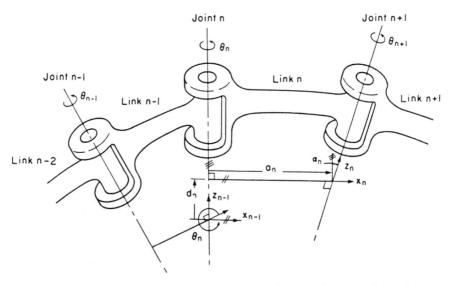

FIGURE 9.2.   Denavit–Hartenberg Parameter Definitions for a Revolute Joint

where θ is the variable joint angle and all the other parameters are fixed. For a prismatic joint the variable d is the joint variable and the transformation matrix relating link n to link $n - 1$ is

$$A_{n.prism} = \begin{bmatrix} cos\theta & -sin\theta \, cos\alpha & sin\theta \, sin\alpha & 0 \\ sin\theta & -cos\theta \, cos\alpha & -cos\theta \, sin\alpha & 0 \\ 0 & sin\alpha & cos\alpha & d \\ 0 & 0 & 0 & 1 \end{bmatrix} \qquad (7)$$

These transformations hold for all manipulator joints as long as all parameters are specified according to the Hartenberg–Denavit conventions. For an N-DOF manipulator the position and orientation of the final link can be expressed in terms of the base coordinates by consecutive premultiplications:

$$T_N = A_1 A_2 \ldots A_N \qquad (8)$$

where A_n is the transformation relating link n to link $n - 1$.

A minimum of six degrees of freedom is required for a manipulator to be able to bring its last link (or the end effector attached to it) into any position and any orientation. This can be rationalized by considering three independent joints necessary for positioning the manipulator in the three-dimensional x-, y-, and z-coordinates, and another three joints for orienting the end effector by rotation about the x-, y-, and z-axes of its local coordinate frame. Although the six degrees of freedom allow any practical position and orientation of the end effector they are, of course, limited by the reach of the manipulator and other physical constraints. The end effector is usually assigned a coordinate system with the z-axis pointing away from the last joint, and the x- and y-axes perpendicular to each other and the z-axis. The x, y, and z unit vectors of the end effector are sometimes referred to as the *normal, (n), orientation (o),* and *approach (a)* vectors. The position of the end-effector coordinate system is denoted by the *position* vector (*p*). These vectors are illustrated in Figure 9.3. Rotations of the end effector (with directions assigned according to the right-hand-rule) are frequently referred to as *roll* (rotation about the end-effector's z-axis), *pitch* (rotation about the y-axis), and *yaw* (rotation about the x-axis).

Because the six degrees of freedom are both necessary and sufficient for any arbitrary position and orientation, robotic manipulators are frequently manufactured with 6-DOF capabilities. Note that with more than six degrees of freedom there are an infinite number of joint-variable combinations which can yield the same total transformation. Such a system is, therefore, redundant. For the 6-DOF manipulator, the position and orientation of the end effector in the world coordinates can be expressed in terms of a T_6 matrix, obtained by multiplication of the six individual link-to-link transforms (A_i). Usually the transformation represented by T_6 does not include the fixed transformation relating the end

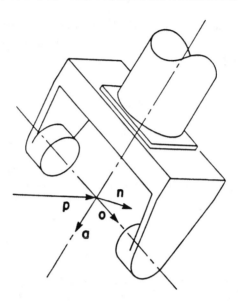

FIGURE 9.3. Position and Orientation of a Robotic End-Effector

effector to the last link of the manipulator and the fixed transform relating the base of the manipulator to the world coordinate system. A general transformation matrix, such as T_6, can be expressed in terms of the *n, o, a,* and *p* vectors of the end effector as:

$$T_6 = \begin{bmatrix} n_x & o_x & a_x & p_x \\ n_y & o_y & a_y & p_y \\ n_z & o_z & a_z & p_z \\ 0 & 0 & 0 & 1 \end{bmatrix} \tag{9}$$

Here T_6 is a function of six joint variables, $\theta_1 - \theta_6$, and, therefore, the position and orientation of the end effector is readily found in terms of the six joint angles (or displacements, in the case of prismatic joints).

Frequently, the inverse of the transformation equation above is of interest, that is, finding $\theta_1 - \theta_6$ (the required joint variables) for a given T_6 matrix (i.e., for a desired position and orientation of the end effector). This requires solving the kinematic Equations (9) for the joint variables. The closed-form solutions have been derived for T_6 (see Paul, [1]). An alternative approach to arriving at these values, by numerical iterations, is frequently applied by commercial robot controllers. Certain positions and orientations of a robot can result in degeneracy. Such singularities arise, for example, when two revolute joint axes are colinear or when two prismatic joint axes are parallel. This is easy to visualize for the case of the parallel prismatic joint axes. Assuming that the joint displacements, d_1 and

d_2, are directed in parallel they will result in a total displacement of $d = d_1 + d_2$. Clearly, the ratio between d_1 and d_2 is not of any consequence, as long as the total adds up to d. Therefore, there are an infinite number of solutions for d_1 and d_2 yielding the desired displacement of d. Similar arguments can be made about the colinear revolute joints of θ_1 and θ_2 adding up to a total revolution of θ. Usually, such singularities are purposely avoided by the robot programmer. Using redundant robot mechanisms, discussed in the next section, is another means of circumventing this problem. Further details of the solution of inverse kinematic equations are not relevant for the succeeding discussion and they are, therefore, omitted here.

A transform graph provides a convenient means of illustrating consecutive transforms of a robotic system. The base of a robotic manipulator is generally specified by a fixed transformation matrix, Z, with respect to the world coordinate system (0). The links of the 6-DOF robot result in an additional transformation $T_6(i) = A_1 A_2 \ldots A_6$, which varies with the stepped movements of the robot (hence, the index i). Finally, the coordinates of the end effector, with respect to the last link of the robot, are obtained from the fixed transformation matrix E. The total transformation matrix, X, is, therefore, obtained as:

$$X(i) = ZT_6(i)E. \tag{10}$$

Transform graphs are frequently employed to describe complicated systems, such as those using multiple manipulators, in a relatively illustrative manner. They will be used here in explaining the weld-path programming algorithms, to be presented later.

Practical programming of robotic manipulators is carried out by specifying a sequence of programmed teach points. The actual trajectory of the end effector consists of straightline segments connected by the teach points. In general, the teach points may be unequally spaced. The programmer may choose to use wide spacing between teach points where the required path trajectory is relatively straight and the orientation of the end effector does not change significantly. For curved trajectory segments, or where the orientation of the end effector goes through abrupt changes, the teach points must be more closely spaced. The teach-point spacing criterion is implied in the fact that the teach points will be connected with straight-lines by the robot controller. For each path segment specified by a pair of adjacent teach points, the robot controller calculates, internally, a number of densely-spaced intermediate points, essentially forming a straightline between the teach points. By moving between these intermediate points, the robot achieves a relatively smooth motion between the teach points. In addition to the teach-point positions and orientations, the programmer can specify a constant speed with which he desires the manipulator to proceed along each teach-point segment. In general, the programmer may request different speeds for different teach-point segments.

In the remainder of this chapter, a robot manipulator of six degrees of freedom and a workpiece positioner of two degrees of freedom will be used as examples for the robot path programming algorithms presented here. Such a system has a total of eight degrees of freedom, and, therefore, it is redundant. In other words, there are an infinite number of joint variable solutions that satisfy any required position and orientation of the end effector. By additionally imposing two independent constraints on the system, the redundancy is eliminated and a unique solution for the eight joint variables can be obtained. For welding applications, the torch and workpiece orientations are used as additional constraints. For nonwelding tasks, similar application-specific constraints can be imposed on the robotic system, and, therefore, the algorithms presented here have a wide applicability beyond the domain of welding. Finally, it should be noted that eight degrees of freedom are not a prerequisite for the presented methodology. An arbitrary system can be constrained by various means to eliminate redundancy. This, and other details of the algorithms, will be discussed in later sections.

3.   REVIEW OF ALTERNATIVE WORK

As stated in the Introduction, the robotic-path programming algorithms presented here are based on applying appropriate constraints to robotic systems with redundant degrees of motion. Furthermore, welding usually requires two mechanisms (the welding manipulator and the positioner) to move in a coordinated manner. The use of coordinated robots with redundant joints in the manipulator-positioner system is, however, by no means new or restricted to welding. Both of these particularities have been tackled by other researchers, separately, and by methods different from those presented here. Although some of the work, briefly reviewed below, is not directly related to robotic welding, these references may be useful for readers pursuing alternative approaches to robotic welding or related applications.

3.1.   Research on Robot Redundancy and Multiple Manipulators

Generally, six degrees of freedom are sufficient for a robot to reach any position and orientation within its reachable domain. Introduction of redundancy by additional joints and links allows further positional and transitional constraint impositions. Thus, redundant manipulators are used to satisfy positional requirements for a number of the manipulator links (e.g., for obstacle avoidance), velocity requirements, and individual joint-range restrictions. Furthermore, degeneracies in the manipulator can be avoided by using redundant joints.

Considerable research has been carried out to exploit the use of redundant robots for added flexibility. A primary advantage of such robots is avoiding collisions with obstacles in the workspace. Collision avoidance is achieved by taking two measures: (a) finding a collision-free path for the work object from the desired start location to the end location, and (b) finding a sequence of joint values moving the robot end effector from the start to the end location while keeping the robot links sufficiently far from the workspace obstacles. The first of these is the path finding problem, addressed by Brooks [5] and Lozano–Perez [6]. The second problem, which essentially requires restrictions on link positions, involves redundant manipulators more directly. Kobrinski and Kobrinski [7] determined certain conditions for the increments of joint coordinates to avoid collisions, while Renaud [8], and Khatib and Le Maitre [9] proposed imposing potentials and force fields around the workspace obstacles to force the robot links away from these objects. Kircanski and Vukobratovic [10] added redundant degrees of freedom to a manipulator, and constrained these with performance criteria that took into account the vicinity of obstacles. The complexity of their algorithms was modest and, thus, they were relatively suitable for real-time implementation. The use of redundancy to avoid singular degenerate configurations of a robot has been studied by Stanisic and Pennock [11]. A nondegenerate kinematic solution was developed for a seven-jointed manipulator. The resulting joint solutions were presented in closed-form equations. A more general solution for the inverse kinematics of redundant robot mechanisms has been developed by Chang [12]. This solution was obtained using the Lagrangian multiplier method and assumed collision avoidance and/or singularity avoidance criteria as additional constraints.

Robotic systems employing multiple manipulators in a concurrent fashion are applicable for a variety of tasks. The welding robot and the workpiece positioner can be regarded as an example of such systems. In general, multiple manipulator systems are needed where skillful handling of complicated and dexterous tasks is required. The main problems introduced by coordinated use of several robots are: (a) potential collisions or physical conflicts between the robots, and (b) programming the robots so that the resulting movements will be coordinated as intended. A recent report on a workshop sponsored by the National Science Foundation (NSF) [13] summarizes the current status of research on coordinated multiple manipulators (CMM) and recommends future research directions. To make CMMs more practical for general industrial applications, considerable research is needed for motion planning, obstacle avoidance, sensors, dynamic modeling, and control strategies. Lee and Mirchandani [14] studied routing and sequencing in a two-manipulator manufacturing cell. An approach was presented which allowed in-process rescheduling of the two robots. Luh and Zheng [15] viewed the CMM problem in a somewhat similar way to the one to be presented here by recognizing the two manipulators and the workpiece as a closed kinematic chain (i.e., transform graph) composed of three sections. They then took the approach of assigning the major part of the movement task to one of the manipulators, the *leader,* while the other one played the role of a *follower,* adjusting its position and orientation so as to keep the kinematic chain closed at all times.

The work of Fernandez and Cook [16] has directly addressed the problem of coordinated multiple manipulators applied for welding. Much of the techniques presented in the remainder of this chapter are based on their approach, with some augmentations. Further discussions of these methodologies are reserved for a later subsection.

3.2.   Current Programming Techniques and Hardware

In the robotic work-cell, the workpiece to be welded is held by a positioner, which usually has a limited degree of motional freedom (typically one to three degrees of freedom). The welding torch is held by the manipulator (which typically has six degrees of freedom). The positioner and the robot are located with respect to each other so that the torch can easily reach the welding seam at all times. This places restrictions on the positions of these two manipulators. The workpiece is accurately placed on the positioner so it can be remounted, or other identical workpieces can be mounted, in exactly the same location at later times.

3.2.1.   Manual Programming.

Using a teach pendant, the operator can manually move the robot and the positioner into any position and orientation within their range. The operator programs each point of the welding sequence by moving both the positioner and the robot into the desired position and orientation, and then all joint variables are stored in the controller memory. Each point involves considerable trial-and-error maneuvers, as several constraints have to be satisfied. The electrode tip has to be brought within a certain range from the weld-seam, and the workpiece has to be appropriately oriented for each point. The angle between the electrode and the seam-tangent, at the programmed point, has to be set at a certain value, and the wire-feeder has to be correctly oriented with respect to the seam. Furthermore, the programmed point locations have to be programmed with the required robot speed, so that the speed of the welding electrode along the seam can be predicted and controlled. Repeating this with the robot pendant for the numerous teach points, constituting a typical weld, is, therefore, a tedious task and prone to erroneous programming. Finally, after completing these off-line programming procedures, the sequence is tested by running both the positioner and the robot through the programmed sequence. The programmed points may have to be repeatedly modified until acceptable results are obtained. By using the algorithms presented later in this article, the operator can substantially simplify the tedious off-line programming procedure and, thus, shorten the programming time.

3.2.2.   CAD/CAM Programming.

A more sophisticated version of the work-cell, discussed above, allows a designer to draft a workpiece on a computer simulation screen and transfer the physical dimensions electronically to manufacturing facilities. Such integration of designing and manufacturing is, frequently, found in *computer integrated manufacturing* (CIM) installations. In this case, manual programming, using teach pendants, is unnecessary as all dimensional data for the workpiece are available in the CIM database. Transforming the available information into a robot-positioner joint parameter sequence, while still satisfying all the required constraints, is by no means trivial in the CIM environment either, however. Using the CIM system, it is a relatively easy matter to establish the teach points with the desired torch orientation. This information can be downloaded to the robot and positioner controllers. This alone does not solve the problem of maintaining constant (or otherwise controlled) travel speed along the weld path, however. Again, the algorithms presented here allow the user to transform this knowledge directly into robot and positioner data, which simultaneously fulfills all required auxiliary constraints.

Simulation of the desired welding sequence before it is implemented in the physical work-cell is practically indispensable. A model of the work-cell (robot, positioner, workpiece, critical obstacles, etc.) may be created in a graphic simulation package. The robot and positioner coordinate sequences (in terms of world coordinates or joint values) are loaded into the simulation computer, where the programmed execution is demonstrated on the screen. The operator verifies that all welding constraints are satisfied and then the sequence can be initiated with the physical robot and the positioner.

4.   DOWNHAND WELDING ALGORITHMS

In this section most of the theory of the off-line programming algorithm is presented, as well as the algorithm itself. The general constraints and requirements of typical arc welding processes are discussed in Section 4.1. Section 4.2 introduces the methodologies on which the proposed algorithm is based, and specifies the components needed to solve the general problems. Finally, these components are combined into the unified algorithm presented in Section 4.3.

4.1.   Welding Constraints

Welding requires a number of constraints to be imposed on the welding robot and the positioner. These constraints differ among the various welding processes. In this section, these constraints will be detailed for some specific cases. The welding constraints for two welding processes, Gas Tungsten Arc Welding (GTAW) and Variable Polarity Plasma Arc Welding (VPPAW) [17], will be used as examples of two different types of robot constraints.

In the GTAW process, an electrical arc is maintained between a nonconsumable tungsten electrode and the workpiece (see Figure 9.4). The arc melts and

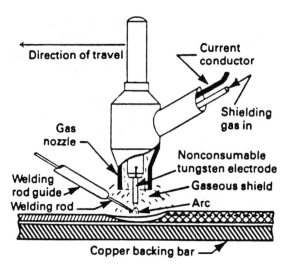

FIGURE 9.4.   Gas Tungsten Arc Welding (GTAW)

fuses the metal being welded, as well as the optional filler metal. The filler metal, if used, is usually fed into the molten weld pool at a controlled rate (wire-feed speed). To isolate the fusion region from the surrounding atmosphere, inert gas (such as argon or helium) is routed through the welding torch, enclosing the electrode, the arc, and the welding pool. Simultaneously, the welding torch moves with a controlled speed along the joint to be welded. To allow the molten metal to settle properly in the joint, the workpiece surface at the fusion point is usually maintained in a horizontal orientation at all times. This is, particularly for manual welding, not a general requirement, but a desired condition for best results. The electrode is either oriented vertically or slightly tilted at a fixed angle from the vertical position, away from the travel direction. Finally, the feed-wire is interjected from the front of the moving electrode, elevated at a fixed angle from the workpiece surface. Thus, the tangent to the weld trajectory at the point of fusion, the axis along the electrode, and the axis along the filler wire should all be in the same vertical plane at all times. In addition to these positional and orientational requirements, the torch must travel at a controllable speed with respect to the weld joint. This controllable speed is usually set constant through-out the entire weld pass.

The hardware for the VPPA welding process is somewhat related to that of the GTAW equipment. The basic elements of the plasma arc torch are the tungsten electrode and the orifice. A relatively minuscule flow of an inert gas (e.g., argon) is guided through the orifice to form the arc plasma. The arc may either be sustained between the electrode and the workpiece (transferred arc), or between the electrode and the constricting nozzle (nontransferred arc). *Keyhole welding*, a specific condition of VPPAW, is achieved with certain combinations of base metal thicknesses, gas flow rates, currents, and torch travel speeds. The weld-pool is relatively small and contains a hole penetrating completely through the base metal. Manual VPPAW, like the GTAW process, can be carried out in most welding positions. Automatic VPPAW is more limited in terms of possible welding positions, and keyhole welding is frequently preferred with the torch always traveling vertically up along the seam.

4.2.   Weld-Path Programming—The Proposed Techniques

The downhand welding position, frequently required for the GTAW process, will be used here for demonstrating the off-line programming algorithm. The downhand welding requirements will, therefore, be implied in the remainder of this chapter, unless otherwise indicated. The requirements for other welding processes (such as VPPAW) can easily be incorporated into the proposed algorithm in the same manner as the downhand constraints are implemented.

The proposed weld-path programming techniques are most clearly outlined by reference to the transform graph illustrated in Figure 9.5. The position and the orientation of the robot (or, more precisely, a coordinate system attached to the robot base), with respect to the world coordinate system, is specified by the fixed transformation matrix Z_R. This allows the system-user to select the world coordinates arbitrarily. The total transformation relating the final link of the robot to its base coordinate system is specified by the variable matrix $T_R(i)$, which changes as the robot moves along the programmed points designated by the index i. For each program point the elements of T_R are functions of one or more joint variables. Here we assume that the joint variables of the robot are θ_{1R}, θ_{2R}, . . . , θ_{6R}. The welding torch, or end effector, is mounted on the end link of the robot. The location of a coordinate system attached to the torch, with respect to the end link, is specified by the fixed transformation matrix E. For the torch to track the welding-seam, and be properly oriented with respect to the seam, the total

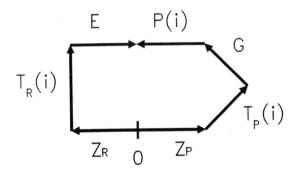

FIGURE 9.5.   The Welding Robot-Positioner Transform Graph with Constraints Satisfied

transformation Z_R $T_R(i)E$ has to be appropriately specified for each programmed point, i.

The positioner side of the transform graph is specified similarly to the robot side. The positioner location in the world coordinates is specified by the fixed transformation Z_P. The location of a coordinate system attached to the positioner-mounting frame with respect to the positioner-base coordinates is $T_P(i)$. For the positioner, we assume only two joint variables, θ_{1P} and θ_{2P}. Each workpiece, or part, mounted on the positioner is assigned a coordinate system, which is fixed with respect to the part. The transformation from the positioner-mounting frame to the part coordinate system is specified by the fixed transformation G. Now that a fixed-coordinate system has been defined with respect to the workpiece, the trajectory of the welded seam is specified. To close the transformation graph, the transformation, $P(i)$ (programmed point i on workpiece with respect to part frame), is specified so that the welding torch coordinates coincide with the coordinates of the programmed point on the workpiece:

$$Z_R \ T_R(i)E = Z_P \ T_P(i)G \ P(i) \qquad (11)$$

for all programmed points, i.

The purpose of the algorithm is to calculate the eight joint variables, θ_{1P}, θ_{2P}, and θ_{1R}, θ_{2R}, $-\theta_{6R}$, for each programmed point, so that the following four requirements are fulfilled at all times:

1. *Torch position control*—The torch tip has to track the weld-seam throughout the entire program point sequence.
2. *Orientation control for downhand welding*—The torch has to be appropriately oriented with respect to the weld tangent (i.e., the tangent to the workpiece surface at the point of welding) at each programmed point, and, simultaneously, the part should be continuously oriented so that the weld tangent is horizontal.
3. *Wire feed orientation control*—The wire-feed contact tube is to be properly oriented in front of the moving torch throughout the entire weld sequence.
4. *Speed control*—The speed of the torch tip, relative to the part surface, is to be controllable. A special case of this is the constant speed requirement normally desired.

One way of applying the algorithm in practice is to program the weld-seam trajectory (the transformation $P(i)$) point by point. This can be done without having to satisfy the four requirements above. After this, only the fixed transformations (Z_R, E, Z_P, G) and the path specification $P(i)$ are known, while the variable $T_R(i)$ and $T_P(i)$ matrices may have arbitrary values. The algorithm obtains solutions for the eight joint variables, θ_{1P} and θ_{2P}, and θ_{1R}, θ_{2R}, $-\theta_{6R}$, (or, equivalently, solves for $T_R(i)$ and $T_P(i)$) by iteratively adjusting the joint variables until requirements 1 through 3 are met. Speed control (4) is achieved by auxiliary calculations. Details of how the above welding requirements are achieved are given in the following subsections.

4.2.1. Torch Position Control.

Before the algorithm is applied, the two sides of the transform graph are, generally, not identical (because T_R and T_P are not yet determined) and the difference is represented by an error transformation *Err:*

$$Err = E^{-1}T_R^{-1}(i)Z_R^{-2}Z_P\,T_P(i)G\,P(i) \tag{12}$$

as shown in Figure 9.6. The error transformation *Err* is, therefore, the transformation between the desired torch location (position and orientation in the part frame) to be reached by the algorithm and the actual initial location. As all transformations, the error *Err* can be represented as successive operations of rotations and translations. Thus, *Err* can be written as:

$$Err = Trans(dx,dy,dz)\,Rot(d\phi_x)\,Rot(d\phi_y)\,Rot(d\phi_z) \tag{13}$$

(i.e., a series of rotations about each of the world coordinate axes followed by a translation). The right-hand side of Equation (13) can be evaluated as a single 4×4 matrix by multiplications of the individual rotation and translation matrices. The resulting matrix will contain elements expressed by sines and cosines of $d\phi_x$, $d\phi_y$, and $d\phi_z$, and dx, dy, and dz. Observe that the notations in Equation (13) suggest differential offsets. Generally, the error transformation may be relatively large when the algorithm is initiated, but it decreases as the algorithm calculations proceed. It turns out that first-degree approximations to the sine and cosine functions of the *Err* matrix, that is, $sin(\phi) \simeq \phi$ and $cos(\phi) \simeq 1$, for small

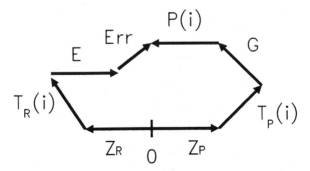

FIGURE 9.6.   The Welding-Robot Positioner Transform Graph with Constraints not Satisfied (*Err* is the Difference Between Actual and Desired Locations.)

ϕ, are sufficient for convergence of the algorithm. The resulting *Err* matrix can, therefore, be expressed as:

$$Err = \begin{bmatrix} 1 & -d\phi_z & d\phi_y & dx \\ d\phi_z & 1 & -d\phi_x & dy \\ -d\phi_y & d\phi_x & 1 & dz \\ 0 & 0 & 0 & 1 \end{bmatrix}. \tag{14}$$

Any vector u transformed by *Err* can be regarded as a sum of three terms, its original location, the error translation, and the error rotation:

$$Err\ u = I\ u + u_{err} + \phi_{err} \times u \tag{15}$$

where I is the 4×4 identity matrix and

$$u_{err} = \begin{bmatrix} dx \\ dy \\ dz \\ 1 \end{bmatrix}; \quad \phi_{err} = \begin{bmatrix} d\phi_x \\ d\phi_y \\ d\phi_z \\ 0 \end{bmatrix}. \tag{16}$$

When the error displacement (u_{err}) and the error rotation (ϕ_{err}) vanish, *Err* becomes the identity matrix. This implies that the torch has been brought into the desired position and orientation with respect to the workpiece, as specified by each path matrix $P(i)$. Note, reducing *Err* down to the identity matrix guarantees nothing more than correct torch location with respect to the part frame. For example, this step alone does not guarantee proper orientation of the workpiece for downhand welding, proper wire-feed orientation, or controlled speed. Those requirements will be treated later.

Before the torch position control part of the algorithm is implemented, the eight joint variables of the system, $\theta_{1P}, \theta_{2P}, \theta_{1R}, \theta_{2R}, \ldots, \theta_{6R}$ have, generally, some arbitrary values, and *Err* is not the identity matrix. For convenience, these eight variables are denoted by a single vector, θ:

$$\theta = \begin{bmatrix} \theta_{2P} \\ \theta_{1P} \\ \theta_{1R} \\ \theta_{2R} \\ \theta_{3R} \\ \theta_{4R} \\ \theta_{5R} \\ \theta_{6R} \end{bmatrix} = \begin{bmatrix} \theta_1 \\ \theta_2 \\ \theta_3 \\ \theta_4 \\ \theta_5 \\ \theta_6 \\ \theta_7 \\ \theta_8 \end{bmatrix}. \tag{17}$$

Recall from Equation (12) that *Err* is a function of $T_P(i)$ and $T_R(i)$, which, in turn, are functions of the eight joint variables. Therefore, θ has to be determined so that *Err* reduces to the identity matrix. Doing this iteratively requires a forcing function for calculating each succeeding θ vector from its predecessor. Such a function is most readily obtained by concatenating u_{err} and ϕ_{err}. Furthermore, because the torch position control is related to the path, $P(i)$, which is defined in the part frame, the forcing function is transformed to the part-reference frame as well. Finally, observing that the fourth components of u_{err} and ϕ_{err} are constants, and of no significance in the iterative recalculations, they can be eliminated through multiplication by a truncated 3×4 identity matrix. The resulting forcing function, $f6_{p-err}$, is, therefore,

$$f6_{p\text{-err}} = \begin{bmatrix} Q \, Act \, u_{err} \\ Q \, Act \, \phi_{err} \end{bmatrix} \tag{18}$$

where *Act,* and Q are defined as:

$$Act = G^{-1}T_P(i)^{-1}Z_P^{-1}Z_R \, T_R(i)E \tag{19}$$

$$Q = \begin{bmatrix} 1 & 0 & 0 & 0 \\ 0 & 1 & 0 & 0 \\ 0 & 0 & 1 & 0 \end{bmatrix}. \tag{20}$$

The forcing function, $f6_{p-err}$, is a six element vector and a function of the eight joint variables. All elements of this vector reach zero when the *Err* becomes the identity matrix. An equivalent criteria is that the norm of $f6_{p-err}$ becomes zero, that is,

$$[f6_{p-err}]^T[f6_{p-err}] \to 0 \qquad \text{as} \qquad Err \to 1 \tag{21}$$

Because $f6_{p-err}$ is a function of θ, this equation can be solved iteratively for θ. Whitney [18] has used Newton's method for this purpose. Finding a zero for an arbitrary function, $f(x)$, of one variable is a common application for Newton's method. A new value for x is iteratively calculated and tried until the corresponding function value is reduced below a specified threshold:

$$x_{j+1} = x_j - [1/f'(x_j)]f(x_j) \tag{22}$$

until

$$f^2(x) < \epsilon \tag{23}$$

In this one-dimensional case $f'(x_j)$ is the first derivative of $f(x)$, evaluated at the jth iteration estimate of x, and ϵ, the convergence criterion, is a small constant. This approach can be extended to functions of more than one variables, such as $f6_{p-err}$. In that case, the iteration algorithm becomes:

$$\theta_{j+1} = \theta_j - [\partial/\partial\theta f6_{p-\text{err}}(\theta_j)]^{-1} f6_{p-\text{err}}(\theta_j) \qquad (24)$$

until

$$[f6_{p-\text{err}}(\theta)]^T[f6_{p-\text{err}}(\theta)] < \epsilon \qquad (25)$$

Differentiation of the $f6_{p-\text{err}}(\theta)$ matrix yields its Jacobian, which relates joint-angle changes to differential torch displacement and rotation variations. In this case, the Jacobian, J, is a 6×8 matrix of the following form:

$$J = \begin{bmatrix} \partial x/\partial\theta_1 & \cdots & \partial x/\partial\theta_8 \\ \partial y/\partial\theta_1 & \cdots & \partial x/\partial\theta_8 \\ \partial z/\partial\theta_1 & \cdots & \partial x/\partial\theta_8 \\ \partial\phi_x/\partial\theta_1 & \cdots & \partial\phi_x/\partial\theta_8 \\ \partial\phi_y/\partial\theta_1 & \cdots & \partial\phi_y/\partial\theta_8 \\ \partial\phi_z/\partial\theta_1 & \cdots & \partial\phi_z/\partial\theta_8 \end{bmatrix} \qquad (26)$$

(e.g., $J_{2,3}$ is the differential of the torch y-coordinate in the part frame with respect to the second joint in the system (θ_{1P}, as defined in Equation 17)). $\partial\phi_x$ denotes a differential rotation about the x-axis, and similar notations hold for rotations about the y- and z-axes. Each column of the Jacobian matrix is composed of three translational derivatives (v) followed by three rotational derivatives (w), that is:

$$J = \begin{bmatrix} v_1 & v_2 & \cdots & v_8 \\ w_1 & w_2 & \cdots & w_8 \end{bmatrix} \qquad (27)$$

The Jacobian for any robotic mechanism is readily calculated as long as all link coordinates are assigned according to the Denavit-Hartenberg conventions, which were discussed earlier. This implies that each link rotates around the z-axis of its local coordinate system. When the Jacobian is analyzed one should keep in mind that only one joint is moved at a time and the rest of the mechanism, connected to each side of that moving joint, can be regarded as two rigid bodies. A unit z-vector (w) is crossmultiplied by the transform-graph segment from the joint being evaluated and back through the torch tip. The result of this is multiplied by the transform from the evaluated joint to the part frame. In this manner, each translational element of the Jacobian is calculated in terms of the individual-link transformations and shown in Equation (28). Similarly, the rotational elements of the Jacobian can be found in terms of the known transformations of the system, and they are shown in Equation (29). Each rotational derivative indicates how much the torch will be rotated around a given part coordinate axis when a specified joint of the mechanism is rotated an incremental angle. Assuming that the positioner and the robots have link transformations A_{1P}, A_{2P}, and A_{1R}, $\cdots$ A_{6R}

counted from their bases, respectively, the elements of the system Jacobian are therefore:

$$v_1 = -QG^{-1}A_{2P}^{-1}[w \times (A_{1P}^{-1}Z_P^{-1}Z_R T_R(i)Er)]$$
$$v_2 = -QG^{-1}A_{2P}^{-1}A_{1P}^{-1}[w \times (Z_P^{-1}Z_R T_R(i)Er)]$$
$$v_3 = QG^{-1}T_P^{-1}(i)Z_P^{-1}Z_R[w \times (A_{1R}A_{2R}A_{3R}A_{4R}A_{5R}A_{6R}Er)] \qquad (28)$$
$$\cdots$$
$$v_8 = QG^{-1}T_P^{-1}(i)Z_P^{-1}Z_R A_{1R}A_{2R}A_{3R}A_{4R}A_{5R}[w \times (A_{6R}Er)]$$

and

$$w_1 = -QG^{-1}A_{2P}^{-1}w$$
$$w_2 = -QG^{-1}A_{2P}^{-1}A_{1P}^{-1}w$$
$$w_3 = QG^{-1}T_P^{-1}(i)Z_P^{-1}Z_R w \qquad (29)$$
$$\cdots$$
$$w_8 = QG^{-1}T_P^{-1}(i)Z_P^{-1}Z_R A_{1R}A_{2R}A_{3R}A_{4R}A_{5R}w$$

where Q, w, and r are:

$$Q = \begin{bmatrix} 1 & 0 & 0 & 0 \\ 0 & 1 & 0 & 0 \\ 0 & 0 & 1 & 0 \end{bmatrix}, \quad w = \begin{bmatrix} 0 \\ 0 \\ 1 \\ 0 \end{bmatrix}, \quad r = \begin{bmatrix} 0 \\ 0 \\ 0 \\ 1 \end{bmatrix} \qquad (30)$$

As before, the 3×4 matrix Q is employed to eliminate the constant terms, which will not be of any use for the weld-programming algorithm. Thus, two rows are eliminated and the resulting Jacobian is a 6×8 matrix.

Returning to the algorithm, the inverse of the derivative in equation (22) is replaced by the pseudoinverse of the Jacobian. For further control of the convergence, the inverted Jacobian may be multiplied by a constant scalar, h, which determines the step size for the iterations. Finally, different weights for the elements of $f6_{p-\text{err}}$ may be preferred in the convergence criterion. This is accomplished by multiplication by a diagonal weighting matrix K. The resulting algorithm for torch position control is, therefore:

$$\theta_{j+1} = \theta_j - h[J^T(JJ^T)^{-1}]f6_{p-\text{err}}(\theta_j) \qquad (31)$$

$$[f6_{p-\text{err}}(\theta)]^T K[f6_{p-\text{err}}(\theta)] < \epsilon \qquad (32)$$

Equations (31) and (32) describe only the algorithm for position and orientation control of the torch with respect to the welded part. Constraints for appropriate workpiece orientation for downhand welding, and wire-feed orientation control, will be discussed in the following subsections.

4.2.2. Downhand Welding Orientation Control

For downhand welding, the workpiece should be oriented so that the part surface at the torch tip is horizontal at all times. To achieve this, the path is examined in the part frame and the desired local vertical-down vector determined at each programmed point. A new forcing function is then developed to align the program point verticals with the gravity gradient in the world coordinates.

Referring to the notations developed earlier in this report, the desired path in the part frame can be represented as:

$$P(i) = [n(i)\ o(i)\ a(i)\ p(i)] \tag{33}$$

where n, o, and a are the 4×1 base vectors of a reference frame attached to the torch and p is the position of the origin of this reference frame with respect to the part frame. A segment of the path, $P(i)$, is depicted in Figure 9.7. The secant i connects point $i - 1$ to point i, and secant $i + 1$ connects points i and $i + 1$. Because most commercial manipulators interpolate movements from one program point to the next one with a straightline, the use of secants rather than curved trajectories is perfectly valid. Furthermore, it is reiterated that it is the

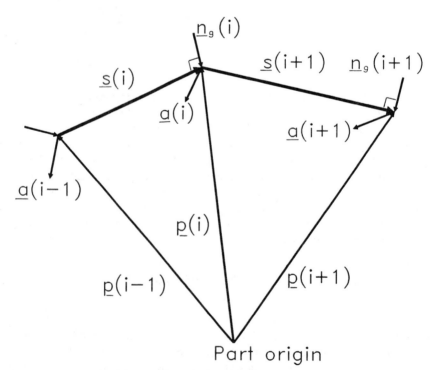

FIGURE 9.7.   Linear Segments between Teach Points along the Weld Path

programmer's responsibility to select the programmed points so that curved trajectories are adequately approximated by straightline segments. A unit vector along each secant is obtained from $P(i)$ as

$$n_s(i) = \frac{p(i) - p(i - 1)}{\|p(i) - p(i - 1)\|} \tag{34}$$

For most welding applications the path secant vector, the approach vector of the torch (which is directed along the plasma flow), and the gravity vector are all in the same vertical plane. Furthermore, as the workpiece surface is to be horizontal at the point of welding, n_g, the desired gravity vector should be perpendicular to the secant $n_s(i)$. This is illustrated in Figure 9.8 and, thus, n_g is calculated as:

$$n_g(i) = \frac{a(i) - [n_s(i) \cdot a(i)]n_s(i)}{\|a(i) - [n_s(i) \cdot a(i)]n_s(i)\|} \tag{35}$$

For vertical-up welding, such as that frequently used for keyhole VPPAW, the n_g vector should always be pointed counter to the travel direction and, thus, it simply becomes:

$$n_{g,\text{vert-up}}(i) = -n_s(i) \tag{36}$$

To achieve downhand position control, the part has to be positioned so that n_g is aligned with g, the Earth's gravity vector, at each programmed point i. If the

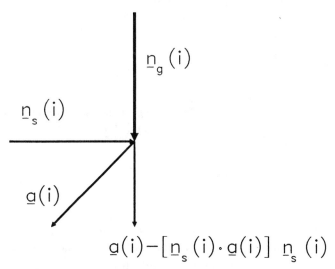

FIGURE 9.8.   Calculation of the Desired Gravity Vector, n_g

coordinate frame for the robot work-cell is chosen so that the x-axis is vertical and pointing up, the gravity vector, g, is $[-1\ 0\ 0\ 0]^T$. Now a forcing function, f_g, bringing n_g into g can be defined as:

$$f_g = n_g(i) \times g \qquad (37)$$

which vanishes as n_g becomes aligned with g:

$$\|f_g\| \to 0 \qquad \text{as} \qquad n_g \to g \qquad (38)$$

Because g is fixed and oriented along the world coordinate x-axis, the f_g forcing vector of Equation (37) will always lie in the horizontal y-z plane. Therefore, the x-component of f_g is always zero and can be eliminated by a truncated identity matrix, as was done in the torch position control section. Using R for this elimination:

$$R = \begin{bmatrix} 0 & 1 & 0 & 0 \\ 0 & 0 & 1 & 0 \end{bmatrix} \qquad (39)$$

the reduced forcing function becomes:

$$f2_{g-\text{err}} = R([Z_P\ A_{1P}\ A_{2P}\ G\ n_g(i)] \times g). \qquad (40)$$

As before, the norm of $f2_{g-\text{err}}$ vanishes as $n_g \to g$.

By augmenting the torch position error function and the downhand welding error function a single error forcing function, $f8_{t-\text{err}}$, for torch position and downhand welding control is obtained:

$$f8_{t-\text{err}} = \begin{bmatrix} f6_{\text{p-err}} \\ f2_{\text{g-err}} \end{bmatrix} \qquad (41)$$

and this error vector becomes 0 when $n_g \to g$. In the same fashion as before the θ vector, which consists of all joint angles in this fully determined system, can be solved by Newton's method:

$$\theta_{j+1} = \theta_j - h\ J^{-1}\ f8_{t-\text{err}}(\theta_j) \qquad (42)$$

$$[f8_{t-\text{err}}(\theta)]^T\ K[f8_{t-\text{err}}(\theta)] < \epsilon \qquad (43)$$

Because of the constraint augmentation the Jacobian is now an 8×8 matrix and, thus, the actual inverse can be used instead of the pseudoinverse used for the torch position control alone. The Jacobian is given by:

$$J = \begin{bmatrix} v_1 & v_2 & v_3 & \cdots & v_8 \\ w_1 & w_2 & w_3 & \cdots & w_8 \\ d_1 & d_2 & 0 & \cdots & 0 \end{bmatrix} \tag{44}$$

where d_1 and d_2 are the following two-dimensional vectors:

$$d_1 = R \, Z_P \, A_{1P} \, w \tag{45}$$

$$d_2 = R \, Z_P \, w \tag{46}$$

The iterative algorithm of Equations (42) and (43) can be used to determine the system joint values so that both torch position control and proper orientation of the workpiece for downhand welding are maintained. The remaining constraint, proper wire-feed orientation, will be addressed in the following subsection.

4.2.3.   Wire-Feed Orientation Control

In most welding applications the wire fed into the molten pool is applied from the front of the welding torch as it moves along the seam. In this subsection, this constraint is added to the robotic welding system while maintaining proper torch position and workpiece orientation. However, because the system is already fully determined (eight equations for eight unknown variables) from the previous sections, addition of one more constraint requires either an additional degree of freedom or relaxation of one of the previous constraints. Adding degrees of freedom to existing robotic systems is usually a major task and, therefore, frequently impractical. The latter approach, relaxing an already established constraint, is, therefore, chosen here. The only requirements for torch position and orientation control were that the torch tip be appropriately placed at the welding joint and that the torch axis along the plasma flow (the approach axis) be coplanar with the path tangent and the gravity gradient. Rotation of the torch about the approach axis is irrelevant and, therefore, that degree of freedom will be used for wire-feed control.

Again, a forcing function for aligning the wire-feed orientation is determined. Referring to Figure 9.9, a vector of unit length, $n_w(i)$, is defined along the axis of the desired wire orientation and pointing counter to its feed direction. Other vectors, such as the path secants, desired gravity, etc., are the same as before. The wire-feed vector, $n_w(i)$, is fixed with respect to the welding torch, but it varies in the part frame. The aim of the wire-feed orientation control algorithm is to rotate the last joint of the robot so that $n_w(i)$, $n_s(i)$, and $a(i)$ become coplanar. Furthermore, noting that this condition can be achieved with the wire feeder directly behind the weld as well as in front of the weld, the forcing function has to be designed so that the algorithm converges to feed wire from the front only.

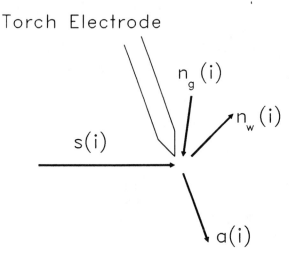

FIGURE 9.9.   Wire-Feed Orientation Illustrated

A scalar forcing function that satisfies the wire-feed orientation constraints defined above is:

$$f1_{w-\text{err}} = [n_w(i) \times n_s(i)] \cdot a(i) \tag{47}$$

In terms of the kinematic matrices of the system, the forcing functions becomes:

$$f1_{w-\text{err}} = [(G^{-1} T_P^{-1}(i)Z_P^{-1} Z_R T_R(i) E n_w(i)) \times n_s(i)] \cdot a(i) \tag{48}$$

To satisfy all of the previously stated constraints, the user of these algorithms determines all eight joint variables using the iterative calculations of Equations (42) and (43). Then one of the eight solutions is revised by the following algorithm:

$$\theta_{6R,j+1} = \theta_{6R,j} + h f1_{w-\text{err}} \tag{49}$$

$$k(f1_{t-\text{err}})^2 < \epsilon. \tag{50}$$

Therefore, the last joint variable of the robot, θ_{6R}, as calculated here, is replaced by the one calculated in Equations (42) and (43). The resulting vector of the eight joint variables, $\theta(i)$, can be determined for each programmed point, i, in the same manner, and the calculated joint values will satisfy the required welding constraints.

4.2.4. Speed Control

Control of the torch speed in the part frame is an important concern for welding. A typical requirement is to keep the speed of the torch tip along the welding trajectory at a specified, constant value. It should be noted that in a multiple manipulator system (e.g., one that consists of a welding robot and a part positioner) the welding speed is not the same as the speed programmed for the robot. The former is specified in the part frame, while the latter is defined with respect to the world frame. A speed of five inches per minute (5 ipm) along the weld seam may be obtained by a stationary torch (in the world frame), while the positioner rotates the workpiece to achieve the desired welding speed. Another scenario might consist of a robot moving the torch with a speed of 25 ipm in the world coordinates and the positioner moving the workpiece so that the weld pool moves in the same direction at 20 ipm, still resulting in a net speed of 5 ipm in the part frame. Because the actual movements of the positioner and the robot are not trivially defined after the downhand welding algorithms have redefined the joint angles, certain measures have to be taken to ensure controlled welding speed.

The original joint angles of the system are transformed by the welding constraint algorithms into a new set of angles for each of the programmed points. As a result, the new $T_R(i)$ and $T_P(i)$ can be found for each programmed teach point along the weld. Assume that the welding speed (i.e., torch speed in the part frame) is specified and assumed to be constant for each segment i, connecting teach points $i - 1$ and i. These desired speeds in the part frame can be designated by the scalars $V_P(i)$, which may vary from one segment to another. For constant speed throughout the entire welding pass all $V_P(i)$ are equal. Now, recall that the programmed teach points in the part frame are defined by a time-varying matrix $P(i)$, which is of the following form:

$$P(i) = \begin{bmatrix} n_{P,x} & o_{P,x} & a_{P,x} & p_{P,x} \\ n_{P,y} & o_{P,y} & a_{P,y} & p_{P,y} \\ n_{P,z} & o_{P,z} & a_{P,z} & p_{P,z} \\ 0 & 0 & 0 & 1 \end{bmatrix} \tag{51}$$

Specifically, the position of each programmed point in the part frame is specified by the fourth column of this matrix and, thus, the distance between any adjacent programmed points is readily found. The time required to traverse a given segment i, $T(i)$, is found as the part-frame distance between points $i - 1$ and i, divided by the welding speed required for segment i:

$$\begin{aligned} T(i) = \surd\{ & [p_{P,x}(i) - p_{P,x}(i - 1)]^2 \\ & + [p_{P,y}(i) - p_{P,y}(i - 1)]^2 \\ & + [p_{P,z}(i) - p_{P,z}(i - 1)]^2 \}/V_P(i) \end{aligned} \tag{52}$$

Now, recall that the position and orientation of the torch tip in world coordinates is given by $Z_R T_P(i)E$. This results in a 4×4 matrix of which the fourth column is the world frame position of the end-effector coordinates:

$$
Z_R T_R(i)E = \begin{bmatrix} n_{R.x} & o_{R.x} & a_{R.x} & p_{R.x} \\ n_{R.y} & o_{R.y} & a_{R.y} & p_{R.y} \\ n_{R.z} & o_{R.z} & a_{R.z} & p_{R.z} \\ 0 & 0 & 0 & 1 \end{bmatrix} \tag{53}
$$

By calculating the segment lengths or distances between successive p-vectors from this equation and dividing them by the time lengths required for each segment, the desired speed of the torch in the world frame, $V_R(i)$, is obtained:

$$
\begin{aligned}
V_R(i) = \surd\{&[p_{R.x}(i) - p_{R.x}(i-1)]^2 \\
+ &[p_{R.y}(i) - p_{R.y}(i-1)]^2 \\
+ &[p_{R.z}(i) - p_{R.z}(i-1)]^2\}/T(i)
\end{aligned} \tag{54}
$$

A more compact form for $V_R(i)$ is

$$
V_R(i) = V_P(i) \frac{\|p_R(i) - p_R(i-1)\|}{\|p_P(i) - p_P(i-1)\|} \tag{55}
$$

which yields the required speed of the end effector in world coordinates for each segment i. These values can be downloaded with the teach points to the actual robot. Note that even when the same welding speed is required throughout all programmed segments (i.e., $V_P(i)$ is constant for all i) the torch speed $V_R(i)$, in the world frame, will, in general, vary due to the interactions between the robot and the part positioner.

This concludes the discussion of the individual algorithms required to achieve torch position control, workpiece orientation for downhand welding, wire-feed orientation, and speed control. To illustrate the generality of the algorithm, the necessary modification for the vertical-up welding constraint was also shown in the subsection on Downhand welding orientation control. In the following section the above discussion is summarized into a united algorithmic framework.

4.3.   Weld Path Programming—The Unified Algorithm

The unified weld-path programming algorithm will now be presented. Refer to Figure 9.10 for the layout of a typical robotic work-cell. The positioner and the robot may be arbitrarily located in the cell and the world coordinate frame of the cell may be arbitrarily located as well. In this case, the x-axis of the world coordinate system is assumed to point straight-up, counter to the gravity vector of

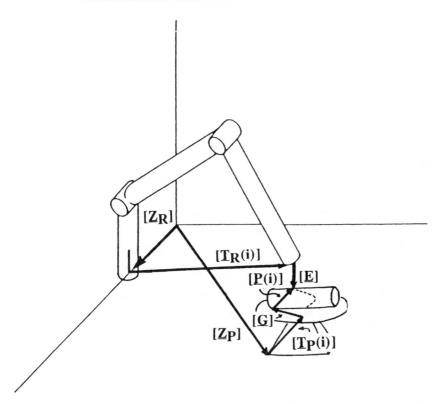

FIGURE 9.10. Work-Cell Definitions for the Weld-Path Programming Algorithm

the Earth. The robot is a 6-DOF mechanism using revolute joints and the positioner is a 2-DOF mechanism, also using revolute joints.

1. *Select World Coordinates*—These can be placed anywhere. Let the x-basis vector point up, opposite to the Earth's gravity vector.
2. *Determine* Z_P *and* Z_R—Assign coordinate systems to the bases of the positioner and the robot (Z_P and Z_R, respectively). These assignments must be according to the Hartenberg–Denavit convention. Z_P and Z_R will be constant 4×4 matrices.
3. *Determine* $T_P(\theta_P)$ *and* $T_R(\theta_R)$—Build a model of the positioner and of the robot, again conforming with the Hartenberg–Denavit rules. This results in two 4×4 matrices, $T_P(\theta_P)$ and $T_R(\theta_R)$, respectively. The elements of T_P contain θ_{1P} and θ_{2P} (the joint angles of the positioner) and the elements of T_R contain $\theta_{1R}, \theta_{2R}, \ldots, \theta_{6R}$ (the joint angles of the robot).
4. *Assign* G, *Determine* E—Assign a local coordinate system to the workpiece. This coordinate system can be selected arbitrarily. G is the transform

from the workpiece coordinates to the last-link coordinate frame of the positioner. Once the part has been permanently attached to the positioner G is constant.

Determine the local coordinate system of the welding torch using the following rules. The origin of this coordinate system is to be at the electrode tip or (accounting for the arc length) on the axis along the plasma flow, about $1/4$ inch in front of the electrode tip. The z-axis (approach vector, a) is oriented along the plasma flow, the x-axis is oriented along the initial torch travel direction, and finally, y is oriented as $z \times x$, conforming with the right-hand rule convention.

5. *Specify P(i)*—Specify the programmed points or teach points along the joint to be welded in terms of the local coordinate system of the workpiece. Each programmed point is to be specified in terms of position as well as orientation, that is, as a complete transformation matrix $P(i)$. The spacing between the points should be selected so that curvatures in the joint trajectory are sufficiently approximated by interpolated straightline segments between adjacent points. The programmed points are to be placed and oriented exactly where the local coordinate system of the torch is desired. Thus, the z-vector should point along the desired plasma flow and the x-vector along the travel direction.

The desired welding path, $P(i)$, can be determined by at least two methods. The more direct approach is possible if the workpiece has been designed and accurately fabricated using a CIM system. In this case, any point on the surface can be accurately determined in terms of position and orientation of the surface at the point. In the absence of such capabilities, the path has to be measured by other means. One way is to move the robot arm through the selected teach points, one at a time, and align the torch on the robot into the appropriate orientation at each point. The path in the part frame can be determined for each point, i, as:

$$P(i) = E^{-1} T_R^{-1}(i) Z_R^{-1} Z_P T_P(i) G(i) \qquad (56)$$

where T_R and T_P can be determined from the joint angles of the robot and the positioner, respectively. Note, this can be done with the positioner joints arbitrarily oriented and they may even be changed throughout this digitizing sequence. An even simpler method is to accurately measure and digitize the workpiece and use the resulting data to determine the desired path.

6. *Select the Next Teach Point from P(i)*—This step is entered once for each teach point along $P(i)$. The first point in the welding sequence is selected first and then Steps 7–9 are iteratively repeated until the $f9_{t-\text{err}}$ error function has been reduced to a small enough vector (and, thus, the joint values of the system have been determined for that teach point).

If all teach points $P(i)$ have been transformed by Steps 7–10, continue to Step 11. At least one iteration of Steps 7–10 is required.

7. *Determine the Total Error Forcing Function—*
   (a) Determine $f6_{p-err}$ (refer to Equation 18):

$$f6_{p\text{-err}} = \begin{bmatrix} Q & Act & u_{err} \\ Q & Act & \phi_{err} \end{bmatrix} \tag{57}$$

where Q is fixed (see Equation 20). Act (Equation 19), u_{err} and ϕ_{err} (Equation 16) will vary from one iteration step to another.
   (b) Determine $f2_{g-err}$ (refer to Equation 39):

$$f2_{g\text{-err}} = R([Z_P \, T_P(i) \, G \, n_g(i)] \times g) \tag{57}$$

where R (Equation 37) and $g = [-1\ 0\ 0\ 0]^T$ are fixed. The vector $n_g(i)$ (Equation 35) is updated for each iteration.
   (c) Determine $f1_{w-err}$ (refer to Equation 46):

$$f1_{w\text{-err}} = [(G^{-1}T_P{}^{-1}(i)Z_P{}^{-1}Z_R T_R(i)E \, n_w(i)) \times n_s(i)] \cdot a(i) \tag{58}$$

where $n_w(i)$ and $n_s(i)$ are updated for each iteration step.
   (d) Form the total error forcing function:

$$f9_{t\text{-err}} = \begin{bmatrix} f6_{p\text{-err}} \\ f2_{g\text{-err}} \\ f1_{w\text{-err}} \end{bmatrix} \tag{59}$$

8. *Determine the Jacobian Matrix (for each program point)—*From Equation (44) the Jacobian is given by:

$$J = \begin{bmatrix} v_1 & v_2 & v_3 & \cdots & v_8 \\ w_1 & w_2 & w_3 & \cdots & w_8 \\ d_1 & d_2 & 0 & \cdots & 0 \end{bmatrix} \tag{60}$$

The individual elements of the Jacobian are calculated in Equations (28), (29), (45), and (46). The Jacobian changes for each iteration.

9. *Determine a New, Iterated Joint Variable Vector—*The new joint variables are found as:

$$\theta_{j+1} = \theta_j + h \begin{bmatrix} J^{-1} & 0 \\ 0 & 1 \end{bmatrix} f9_{t\text{-err}}(\theta_j) \tag{61}$$

where h is appropriately selected for adequate convergence stability. Note, θ is a 9×1 vector. Its first seven elements are A_{2P}, A_{1P}, A_{1R}, A_{2R}, . . . , A_{5R}. θ_8 is the last robot joint variable and its value is discarded, as it is calculated without the wire-feed constraint. Instead, θ_9, the last element of H, is the value used for the last joint variable as it is derived with the wire-feed constraint.

10. *Determine If Convergence Has Been Reached*—A new iterated set of joint values was reached in Step 9 above. For these new values (iteration $j + 1$) calculate if convergence has been reached:

$$[f9_{t-\text{err}}(\theta_{j+1})]^T \, K'[f9_{t-\text{err}}(\theta_{j+1})] < \epsilon \qquad (62)$$

K' is a diagonal weighting matrix for optional weight distribution of the joint variables. Commonly, it is chosen as the 9×9 identity matrix.

   If the inequality (62) holds, convergence has been reached. The vector θ_{j+1} is stored as the new joint values for teach point $P(i)$, which satisfy the downhand constraints. Continue back to Step 6 and select the next teach point, $P(i + 1)$.

   If this inequality does not hold, more iterations are needed for the current teach point $P(i)$. Steps 7–9 must be repeated with θ_{j+1} replacing θ_j.

11. *Calculate $V_R(i)$*—Now that all joint angles of the path have been recalculated to satisfy the downhand welding requirements, the corresponding speed values for the robot are calculated. The part-frame speed required for each segment i, from teach point $i - 1$ to teach point i, is assumed known in the array $V_P(i)$. For a constant-speed weld all $V_P(i)$ are equal. Then the speed values in the world frame, loaded to the robot, are found as:

$$V_R(i) = V_P(i) \frac{\|p_R(i) - p_R(i\text{-}1)\|}{\|p_P(i) - p_P(i\text{-}1)\|} \qquad (63)$$

where $p_R(i)$ and $p_P(i)$ are found, as shown in Equations (51) and (53), respectively.

   This concludes the formal downhand welding algorithm. At this stage the eight system joint variables have been recalculated for each teach point i along the welding path, so that the downhand welding constraints are satisfied. Changing the constraints for other requirements than those of downhand welding is relatively straightforward and, as an example, changing $n_g(i)$, as shown in Equation (36), is all that is needed to implement vertical-up welding. In addition to the joint angles, the algorithm has calculated the necessary speeds required for the robot to maintain controlled or constant welding speed throughout the weld.

5. SIMULATION RESULTS

To verify and demonstrate the accuracy of the algorithm, computer simulations of it are presented, again using downhand welding constraints to exemplify the general capabilities. The simulations presented here were designed to mimic welding applications carried out at the NASA Materials Processing Laboratories at the Marshall Space Flight Center in Alabama.

The robot is the Cincinnati–Milacron T^3-776. It has six degrees of freedom and all joints are revolute. The positioner is an RP-25 from the Advanced Robotics Corporation. It has two revolute joints, one revolving the mounting plate in its plane and one to tilt it off the horizontal orientation. In addition to these two degrees of freedom, the positioner height can be indexed off-line, but the index parameter is usually assumed fixed for any given application.

The purpose of the computer simulations is twofold. First, graphical simulations verify if correct positions and orientations of the welding torch and the workpiece are achieved for all teach points along the programmed welds. Secondly, constant welding speed capabilities with respect to the part frame are demonstrated. For the simulations, ROBOSIM, a graphical simulation package developed by Fernandez [19], was employed in conjunction with FORTRAN programs, used for implementation of the robot weld-path programming algorithm. Most of the simulations were run on a VAX-8800 computer, except for some of the final graphical displays, which were demonstrated on a Hewlett–Packard 9000/350-SRX workstation.

5.1. Graphical Simulations

Having assigned a coordinate frame to the simulated welding torch, its desired position and orientation, in the part frame, is determined for each teach point. The workpiece selected for the simulations is a nozzle, which is used in the NASA Space Shuttle Main Engine. The nozzle is welded along its axis of symmetry from one end to the other, following the surface curvatures along the way. The coordinate frame assigned to the workpiece is located at the starting point of the weld. The teach-point coordinates (position and orientation) for the nozzle weld are listed in Table 9.1. This kind of data, or other data which essentially contains the same information, are necessary to specify the welding path for the workpiece. For example, the six joint values from a manually guided welding robot can be transformed into the form, shown in Table 9.1. Each straightline segment of the nozzle weld is specified by its endpoints. Recalling that the straightline segments are joined by curved arcs, each arc is assigned three teach points in addition to its endpoints, shared by the connecting lines. The resulting teach points are fairly unevenly spaced, which is somewhat representative of actual teach-point assignments for this workpiece. The nozzle teach points are indicated in Figure 9.11.

TABLE 9.1.   Path definition for the nozzle weld

Point No:	X-pos. [in]:	Y-pos. [in]:	Z-pos. [in]:	Roll [degr]:	Pitch [degr]:	Yaw [degr]:
1	−0.000	0.000	0.000	0.000	−125.538	0.000
2	−3.571	0.000	5.000	0.000	−125.097	0.000
3	−3.642	0.000	5.100	0.000	−111.172	0.000
4	−3.680	0.000	5.200	0.000	−95.465	0.000
5	−3.690	0.000	5.300	0.000	−89.021	0.000
6	−3.671	0.000	6.400	0.000	−86.248	0.000
7	−3.648	0.000	6.750	0.000	−79.652	0.000
8	−3.584	0.000	7.100	0.000	−72.956	0.000
9	−3.477	0.000	7.450	0.000	−65.758	0.000
10	−3.319	0.000	7.800	0.000	−64.099	0.000
11	−0.940	0.000	12.700	0.000	−68.653	0.000
12	−0.627	0.000	13.500	0.000	−71.200	0.000
13	−0.389	0.000	14.200	0.000	−73.706	0.000
14	−0.155	0.000	15.000	0.000	−76.076	0.000
15	0.043	0.000	15.800	0.000	−82.176	0.000
16	0.414	0.000	18.500	0.000	−82.176	0.000

Each of the 16 teach points is specified in terms of its x-, y-, and z-coordinates, and roll, pitch, and yaw (rotations about the z-, y-, and x-axes, respectively). These parameters are all defined with respect to the part-coordinate frame.

Snapshots from the nozzle-welding simulation, after the downhand welding algorithm is applied, are shown in Figures 9.12, 9.13, 9.14, and 9.15. It should be noted that the workpiece positioner and the welding manipulator must move simultaneously and in a coordinated manner to maintain the downhand welding conditions.

5.2.   Constant-Speed Verification

Although the position and orientation of both the workpiece and the welding torch are clearly demonstrated in the graphical simulation, they do not convey torch travel-speed with respect to the workpiece. The speed of the simulation on the screen is largely related to the calculation speed for graphics updates and other factors of the simulation program, and, therefore, the simulation speeds are not a reliable indicator of the speed of the actual physical system. Therefore, the speed of the torch in the world coordinates, as well as in the part coordinates, is explicitly calculated for each teach-point interval. The world coordinate speed of the torch is calculated using Equation (55). The distances between teach points in the part coordinate frame are found by recalculating:

$$P'(i) = E^{-1}T_R'{}^{-1}(i)Z_R{}^{-1}Z_P T_P'(i)G \qquad (64)$$

Nozzle Cross Section
– Center Axis at X=0 –

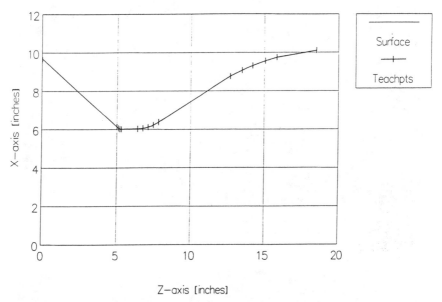

FIGURE 9.11.   A Half Cross-Section of the Nozzle, with the Teach Points Designated with Vertical Marks

where $T_i'(i)$ and $T_p'(i)$ contains the system joint values *after* the downhand algorithm has been applied. If the new joint values satisfy all of the constraints the new $P'(i)$ is identical to the original $P(i)$, the distances between teach points are unchanged, and, thus, the speed is constant as originally specified.

The results of the speed simulations for the nozzle weld are shown graphically in Figure 9.16. A constant welding speed of 6 ipm was chosen for the simulation. The results show the torch speed in the world frame and torch speed in the part frame throughout the weld, calculated as explained above. The torch speed in the part frame varies widely. The sharp peak of up to 120 ipm lasts for the short period, during which the arc passes across the sharpest curvature of the weld. Another, considerably lower peak occurs during the second curvature, and so on. Note, initially the torch speed is 6 ipm, which means that the torch moves at a constant speed along the nozzle-joint, while the workpiece is held stationary. This is what one would expect; initially the workpiece is oriented so that its first linear segment between the teach points is horizontal and, therefore, no further

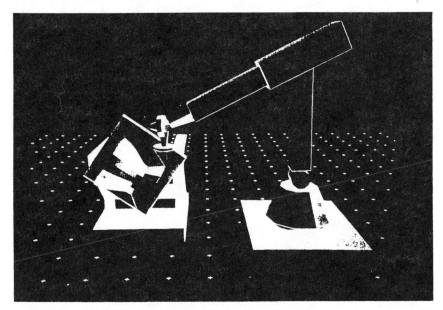

FIGURE 9.12.   A View of the Robotic Work Cell at the Start of the Nozzle Weld

FIGURE 9.13.   Welding along the First Straightline Segment (between Teach Points 1 and 2)

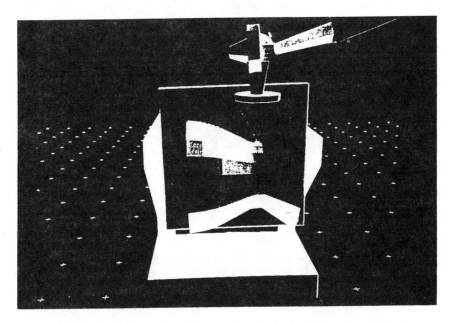

FIGURE 9.14.   Welding along the Second Straightline Segment

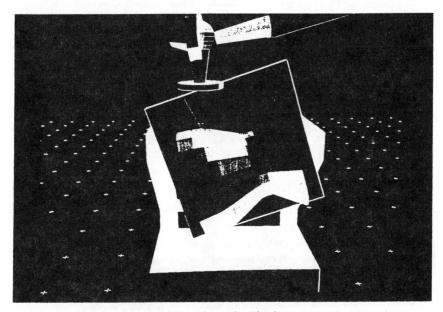

FIGURE 9.15.   Welding along the Third Curvature Segment

Robotic Torch Speed

– Nozzle Weld –

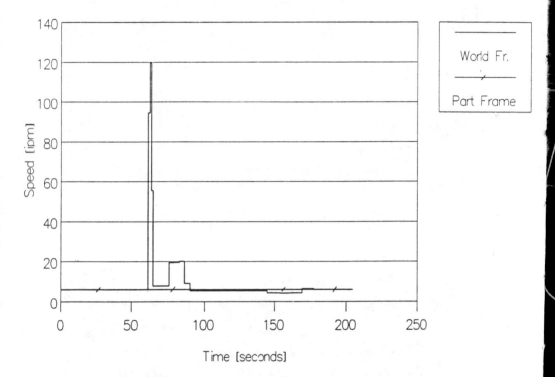

FIGURE 9.16.   World Frame and Part Frame Torch Speeds for the Nozzle Weld

rotations of the workpiece are needed during that segment. The same applies to
the final segment of the weld. The discontinuities in the robot speed curve are
due to the fact that each segment connecting adjacent teach points is assumed to
be a straightline and it is assumed that constant speed is maintained during each
such interval. As mentioned before, it is up to the programmer to select the teach
points; denser selection would result in an overall smoother speed curve. Finally,
it should be noted that the speed with respect to the workpiece is essentially
constant at 6 ipm, as required.

6. FUTURE RESEARCH—CONCLUSIONS

The purpose of the research presented in this chapter was to establish algorithms for programming of robots with multiple and redundant axes of motion, applied to welding. Of particular interest were *torch position control, downhand welding control, wire guide orientation control,* and *welding speed control.*

Substantial time can be saved and programming accuracy can be ensured by application of the algorithm presented here. In the case of programming the weld path by guiding the robot with a teach pendant, the programmer can orient the workpiece by aligning the positioner joints into any convenient position, at the beginning of the programming sequence as well as at any time throughout the programming process. The torch is guided to the first teach point on the workpiece, where the programmer only ensures adequate proximity to the desired teach-point location and the desired torch orientation with respect to the workpiece. Wire-feed orientation and workpiece orientation are no longer relevant. This can be repeated for all teach points. In the case of a CAD/CAM system, the coordinates and desired torch orientations at the selected teach points can be extracted from the workpiece data, from which all information for the algorithm can be derived. In either case, the desired welding speed or speeds are also specified. From this information the proposed welding algorithm calculates the desired positioner and robot joint angles, and robot speeds, required to make the weld with the desired welding constraints. In short, a process which either was impossible or, at best, a series of tedious trial-and-error operations, is turned into an efficient one-pass task using the proposed algorithms.

The basic algorithm, presented here, can be extended so it applies to more general robotic systems than demonstrated in this overview. For example, a framework is being considered which enables the programmer to apply the required welding constraints to various robot and positioner types. For a total number of degrees of freedom other than eight, the system may suggest or apply additional constraints or notify the user if the welding requirements cannot be satisfied with the suggested robotic mechanisms. An extended algorithm framework should also automatically impose the necessary constraints for welding processes other than GTAW and VPPAW, which were demonstrated here. Real-time correction of the welding path, based on joint tracking sensors, may be implemented in the algorithm as well.

REFERENCES

1. R.P. Paul, *Robot Manipulators Mathematics, Programming, And Control,* MIT, Cambridge, MA, 1981.
2. A.J.Critchlow, *Introduction to Robotics,* Macmillan, 1985.
3. J.D. Lane, *International Trends in Manufacturing Technology—Robotic Welding,* IFS, U.K., 1987.

4. J. Denavit, and R.S. Hartenberg, "A Kinematic Notation for Lower-Pair Mechanisms Based on Matrices," *ASME Journal of Applied Mechanics*, Vol. 19, No. 2, June 1955, pp. 215–221.

5. R.A. Brooks, "Solving the Find-Path Problem by Good Representation of Free Space," *IEEE Transactions of Systems, Man and Cybernetics*, Vol. 13, No. 3, March/April 1983, pp. 190–197.

6. T. Lozano-Perez, "Spatial Planning: A Configuration Space Approach," *IEEE Transactions on Computers*, Vol. 32, No. 2, February 1983, pp. 108–120.

7. A.A. Kobrinski, and A.E. Kobrinski, "Manipulation System Trajectory Synthesis in the Environment with Obstacles," *Comm. of USSR Academy of Science*, Vol. 224, No. 6, June 1983, pp. 23–31.

8. M. Renaud, "Contribution a l'etude de la modelisation et de la commande des systemes mechaniques articules," Ph.D. Thesis, Universite de Paul Sabatier, Toulouse, France, December 1975.

9. O. Khatib, and J.F. Le Maitre, "Dynamic Control of Manipulators, Operating in a Complex Environment," *Proc. 3rd International CISM-IFToMM Symposium on Theory and Practice of Robots and Manipulators*, Udine, Italy, 1987.

10. V. Kircanski, and M. Vukobratovic, "Contribution to Control of Redundant Robotic Manipulators in an Environment with Obstacles," *The International Journal of Robotics Research*, Vol. 5, No. 4, Winter 1986, pp. 112–119.

11. M.M. Stanisic, and G.R. Pennock, "A Nondegenerate Kinematic Solution of a Seven-Jointed Robot Manipulator," *The International Journal of Robotics Research*, Vol. 4, No. 2, Summer 1985, pp. 10–20.

12. P.H. Chang, "A Closed-Form Solution for Inverse Kinematics of Robot Manipulators with Redundancy," *IEEE Journal of Robotics and Automation*, Vol. RA-3, No. 5, October 1987, pp. 393–403.

13. A. Koivo, and G.A. Bekey, "Report of Workshop on Coordinated Multiple Robot Manipulators: Planning, Control, and Applications," *IEEE Journal of Robotics and Automation*, Vol. 4, No. 1, February 1988, pp. 91–93.

14. E. Lee, and P.B. Mirchandani, "Concurrent Routing, Sequencing, and Setups for a Two-Machine Flexible Manufacturing Cell," *IEEE Journal of Robotics and Automation*, Vol. 4, No. 3, June 1988, pp. 256–264.

15. J.Y.S. Luh, and Y.F. Zheng, "Constrained Relations between Two Coordinated Industrial Robots for Motion Control," *The International Journal of Robotics Research*, Vol. 6, No. 3, Fall 1987, pp. 60–70.

16. K.R. Fernandez, and G.E. Cook, "A Generalized Method for Automatic Downhand and Wirefeed Control of a Welding Robot and Positioner," Technical Paper 2807, NASA, Washington, DC, February 1988, pp. 1–48.

17. C. Weisman, *Welding Handbook*, Vol. 1–5, Seventh edition, American Welding Society, Miami, FL, 1976.

18. D.E. Whitney, "Resolved Motion Rate Control of Manipulators and Human Prostheses," *IEEE Transactions on Man-Machine Systems*, Vol. MMS-10, No. 2, June 1969, pp. 47–53.

19. K.R. Fernandez, "Robotic Simulation and a Method for Jacobian control of a Redundant Mechanism with Imbedded Constraints," Ph.D. Dissertation, Vanderbilt University, Nashville, TN, May 1988, pp. 22–43.

Author Index

Subject Index

a foot in diameter. At the bottom of the insulator was another corona ring. Imagine a doughnut flat on the table with a pill bottle sitting on the hole and a second doughnut at the other end of the bottle. The remainder of the breaker comprised a mechanical switch to accomplish the permanent switching as the plasma interrupter could only function briefly, but it could do this at very high speed, a few thousandths of a second. Also on the truck bed was a huge resistor to absorb the energy stored in the transmission line when it was opened and of course all the electronics to manipulate the complete device.

For safety reasons the test sequence was written out in detail with command responsibility assigned to the operator of Los Angeles Department of Water and Power (LADWP). His commands were to be instantly obeyed and no actions were to be taken except at his command. The entire test sequence only occupied 2 minutes, but the events were timed out in milliseconds based on a computer simulation of the entire system between The Dalles and Sylmar which Hughes had written and run to convince LADWP that the test would not be destructive. The 2 minutes of test covered three pages of single-spaced typing. The entire crew of both Hughes and LADWP memorized those three pages as there would be no time to "read up" if an emergency occurred during the test.

The computer simulation of the test caused its' own problems. It was so large that it would not run on the Hughes central computer (IBM). We found a Control Data Corp (CDC) until that was available for rental usage. We shipped over the tape, they converted it to IBM cards and began simulations. We ran several different conditions and all looked good, however, I wanted one last one more indicative of the test. CDC called and said they were sorry, but they couldn't run it. They had dropped the boxes of IBM cards and lost the sequence. I offered, "OK, these things happen. Just re-sequence them from the card serial numbers." There was a silence, then, "the cards were not given serial numbers." I exhaled silently and kept my cool, "OK, rerun the tape and create new cards and this time serial number them." There was a longer silence, "We can't find the tape." I hung up before I blew up and called our software writer to ask him to furnish a backup tape to CDC. He said, "They have the only tape in existence." I took ten deep breaths while wondering from what world software people came. It would do no good to blame anyone, the question was the adequacy of the simulations that we already had. A review showed these to be 99% adequate. We reviewed then with LADWP and they approved the test.

The truck was driven to Sylmar and the necessary connections made after a test of function on the Hughes electronics. The day of the test dawned with a moderate rain. The truck and breaker had been installed under a circus tent in anticipation of rain, so we were not worried and gave the Hughes OK to begin the test. On the LADWP signal to begin, high-speed video tape recorders started, LADWP began the countdown for the application of the short circuit. Three hundred million watts of power flowed out of The Dalles and half, or 150 million watts through the DC circuit breaker back to Oregon.

The short would be applied by a man with a bow and arrow. The arrow carried a metal wire connected to ground. He would shoot the arrow over one of the transmission line conductors, being careful not to hit the other. Once the arc was formed when the wire melted the arc would sustain itself as a plasma from the transmission line to ground while our breaker attempted to disconnect that line, having detected the high current of a short circuit. The short was to be applied about 500 miles from Sylmar. Thus, communication was by radio to start the short at the correct time on the test schedule.

Before the archer could loose his arrow, there was a brilliant flash and loud BANG on the truck. This caused the LADWP operator to immediately terminate the sequence. Yes, emergency termination was part of the typed test sequence.

Inspection showed there was some damage to the parts of the breaker. I immediately asked the Hughes project leader to have all damaged parts replaced on a hurry-up basis while he and I went to look at the video tapes to see if we could find out what happened. LADWP asked if we wanted to abandon the test. I replied NO we were replacing damaged parts and would be ready to go again in two hours.

The videos were well done, but at first not informative. There was simply a huge, brilliant arc between the two corona rings on the high voltage connector. We played them again and again with no conclusion as to cause of the arc. Then once more we looked at the video at super-slow motion. Then I saw it. There was a droplet of water in the air just outside the corona rings. The electric field of the DC voltage tore the droplet apart into a long string of vapor which reduced the open insulated space between the rings until an electrical breakdown occurred. Now we knew what, but where had a water droplet come from in that sensitive spot? We went out to the breaker under the tent and looked upward. Indeed, there was a rip in the tent at just the right spot to let a droplet form and fall between the corona rings. The project manager and I let out a whoop of success and started part of the crew repairing the torn tent while we went back to tell LADWP what we found.

They reviewed the tapes, looked at the tent and agreed that the test could proceed. At this time LADWP headquarters called to advise that the test time window would close in 15 minutes and they would have to go back to normal operations. For Hughes there would be an inglorious end to two years of work by a dozen people and who knows when we could get approval from them for another test sequence. We would not finish the replacement of damaged parts for another half hour. I asked the LADWP management for one more hour, citing the importance of the test to LADWP and EPRI. The voice on the phone said, "If we give you that hour, will you agree not to ask for more time? You should know that we save $128,000. every hour that we operate Sylmar as compared our local generation." I swallowed hard and in my best voice said confidently, "I agree not ask for more time." There was a brief silence on the phone followed by, "OK, you have one hour."

In 20 minutes we were finished with parts replacement. After a quick inspection of the breaker and the tent, I said to the LADWP test director, "Hughes is ready for a retest." The test sequence began. At the appropriate point the archer released his arrow, the arc formed on the transmission line and the Hughes switch was heard to operate. The test sequence shutdown was ordered followed by a silence. We waited, perhaps two minutes, but it seemed an hour. Then the LADWP test director who had access to the station records said over the PA, "Congratulations, you did it." The entire Hughes crew and field personnel of LADWP let out a huge whooping cheer. After we had checked the station records, just to be sure, I told my crew, Let's go to the restaurant for a couple beers and dinner. It's on me!"

This was a worlds' first! A circuit breaker, first ever of its' kind had cleared a short circuit on a branched HVDC system without interrupting the energy flow in the un-shorted branch. The accomplishment was recognized the following year when the Institute of Electrical and Electronic Engineers devoted